Basic Computation

Working with Whole Numbers

Loretta M. Taylor, Ed. D.
Mathematics Teacher
Hillsdale High School
San Mateo, California

Harold D. Taylor, Ed. D.
Head, Mathematics Department
Aragon High School
San Mateo, California

DALE
SEYMOUR
PUBLICATIONS
P.O. BOX 10888
PALO ALTO, CA 94303

Editors: Elaine C. Murphy, Mary Heinrich
Production Coordinator: Ruth Cottrell
Cover designer: Michael Rogondino
Compositor: WB Associates
Printer: Malloy Lithographing

ISBN 0-86651-000-1
Order Number DS01181

bcdefghi-MA-8765432

DALE
SEYMOUR
PUBLICATIONS
P.O. BOX 10888
PALO ALTO, CA 94303

ABOUT THE PROGRAM

WHAT IS THE BASIC COMPUTATION LIBRARY?

The books in the BASIC COMPUTATION library together provide comprehensive practice in all the essential computational skills. There are practice books and a test book. The practice books consist of carefully sequenced drill worksheets organized in groups of five. The test book contains daily quizzes (160 quizzes in all), semester tests, and year-end tests written in standardized-test formats.

If you find this book effective, you may want to use others in the series. Build your own library to suit your own needs.

BOOK 1	WORKING WITH WHOLE NUMBERS
BOOK 2	UNDERSTANDING FRACTIONS
BOOK 3	WORKING WITH FRACTIONS
BOOK 4	WORKING WITH DECIMALS
BOOK 5	WORKING WITH PERCENTS
BOOK 6	UNDERSTANDING MEASUREMENT
BOOK 7	FINDING AREA AND PERIMETER
BOOK 8	WORKING WITH CIRCLES AND VOLUME
BOOK 9	APPLYING COMPUTATIONAL SKILLS
TEST BOOK	BASIC COMPUTATION QUIZZES AND TESTS

WHO CAN USE THE BASIC COMPUTATION LIBRARY?

Classroom teachers, substitute teachers, tutors, parents, and persons wishing to study on their own can use these materials. Although written specifically for the general math classroom, books in the BASIC COMPUTATION library can be used with any program requiring carefully sequenced computational practice. The material is appropriate for use with any person, young or old, who has not yet certified computational proficiency. It is especially suitable for middle school, junior high school, and high school students who need to master the essential computational skills necessary for mathematical literacy.

WHAT IS IN THIS BOOK?

This book is a practice book. In addition to these teacher notes, it contains student worksheets, example problems, and a record form.

Worksheets

The worksheets are designed to give even the slowest student a chance to master the essential computational skills. Most worksheets come in five equivalent forms allowing for pretesting, practice, and posttesting on any one skill. Each set of worksheets provides practice in only one or two specific skills and the work progresses in very small steps from one set to the next. Instructions are clear and simple, with handwritten samples of the exercises completed. Ample practice is provided on each page, giving students the opportunity to strengthen their skills. Answers to each problem are included in the back of the book.

Example Problems

Fully-worked examples show how to work each type of exercise. Examples are keyed to the worksheet pages. The example solutions are written in a straightforward manner and are easily understood.

Record Form

A record form is provided to help in recording progress and assessing instructional needs.

Answers

Answers to each problem are included in the back of the book.

HOW CAN THE BASIC COMPUTATION LIBRARY BE USED?

The materials in the BASIC COMPUTATION library can serve as the major skeleton of a skills program or as supplements to any other computational skills program. The large number of worksheets gives a wide variety from which to choose and allows flexibility in structuring a program to meet individual needs. The following suggestions are offered to show how the BASIC COMPUTATION library may be adapted to a particular situation.

Minimal Competency Practice

In various fields and schools, standardized tests are used for entrance, passage from one level to another, and certification of competency or proficiency prior to graduation. The materials in the BASIC COMPUTATION library are particularly well-suited to preparing for any of the various mathematics competency tests, including the mathematics portion of the General Educational Development test (GED) used to certify high school equivalency.

Together, the books in the BASIC COMPUTATION library give practice in all the essential computational skills measured on competency tests. The semester tests and year-end tests from the test book are written in standardized-test formats. These tests can be used as sample minimal competency tests. The worksheets can be used to brush up on skills measured by the competency tests.

Skill Maintenance

Since most worksheets come in five equivalent forms, the computation work can be organized into weekly units as suggested by the following schedule. Day one is for pretesting and introducing a skill. The next three days are for drill and practice followed by a unit test on the fifth day.

AUTHOR'S SUGGESTED TEACHING SCHEDULE

	Day 1	Day 2	Day 3	Day 4	Day 5
Week 1	pages 1 and 2 pages 11 and 12	pages 3 and 4 pages 13 and 14	pages 5 and 6 pages 15 and 16	pages 7 and 8 pages 17 and 18	pages 9 and 10 pages 19 and 20
Week 2	pages 21 and 22 pages 31 and 32	pages 23 and 24 pages 33 and 34	pages 25 and 26 pages 35 and 36	pages 27 and 28 pages 37 and 38	pages 29 and 30 pages 39 and 40
Week 3	pages 41 and 42 pages 51 and 52	pages 43 and 44 pages 53 and 54	pages 45 and 46 pages 55 and 56	pages 47 and 48 pages 57 and 58	pages 49 and 50 pages 59 and 60
Week 4	pages 61 and 62 pages 71 and 72	pages 63 and 64 pages 73 and 74	pages 65 and 66 pages 75 and 76	pages 67 and 68 pages 77 and 78	pages 69 and 70 pages 79 and 80
Week 5	pages 81 and 82 pages 91 and 92	pages 83 and 84 pages 93 and 94	pages 85 and 86 pages 95 and 96	pages 87 and 88 pages 97 and 98	pages 89 and 90 pages 99 and 100

The daily quizzes from BASIC COMPUTATION QUIZZES AND TESTS can be used on the drill and practice days for maintenance of previously-learned skills or diagnosis of skill deficiencies.

A five-day schedule can begin on any day of the week. The author's ideal schedule begins on Thursday, with reteaching on Friday. Monday and Tuesday are for touch-up teaching and individualized instruction. Wednesday is test day.

Supplementary Drill

There are more than 18,000 problems in the BASIC COMPUTATION library. When students need more practice with a given skill, use the appropriate worksheets from the library. They are suitable for classwork or homework practice following the teaching of a specific skill. With five equivalent pages for most worksheets, adequate practice is provided for each essential skill.

HOW ARE MATERIALS PREPARED?

The books are designed so the pages can be easily removed and reproduced by Thermofax, Xerox, or a similar process. For example, a ditto master can be made on a Thermofax for use on a spirit duplicator. Permanent transparencies can be made by processing special transparencies through a Thermofax or Xerox.

Any system will run more smoothly if work is stored in folders. Record forms can be attached to the folders so that either students or teachers can keep records of individual progress. Materials stored in this way are readily available for conferences.

EXAMPLE PROBLEMS

ADDITION OF WHOLE NUMBERS

EXAMPLE 1 Find the sum of 463, 294, and 139.

Solution: 463 + 294 + 139 = 896

EXAMPLE 2 Write the sum of 8 and 7.

Solution: 8 + 7 = 15

SUBTRACTION OF WHOLE NUMBERS

EXAMPLE 1 Subtract 276 from 511.

Solution: 511 − 276 = 235

EXAMPLE 2 Find the difference between 15 and 89.

Solution: 89 − 15 = 74

ADDITION AND SUBTRACTION OF WHOLE NUMBERS

EXAMPLE 1 Use addition and subtraction to complete the chart.

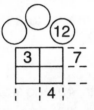
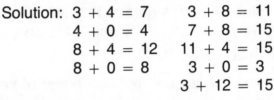
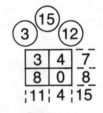

Solution: 3 + 4 = 7   3 + 8 = 11
4 + 0 = 4   7 + 8 = 15
8 + 4 = 12   11 + 4 = 15
8 + 0 = 8   3 + 0 = 3
             3 + 12 = 15

MULTIPLES OF WHOLE NUMBERS

EXAMPLE 1 Write five numbers between 100 and 160 that are multiples of 2.

Solution: If a number is a multiple of 2, then its final digit must be 0, 2, 4, 6, or 8. Thus, five numbers are 114, 120, 136, 142, and 158.

EXAMPLE 2 Write the facts about multiples of 4.

Solution: If the last two digits of a number form a number that is divisible by 4, the number is a multiple of 4. For example: 312 is a multiple of 4 since the last two digits, 12, form a number divisible by 4.

RECOGNIZING MULTIPLES OF WHOLE NUMBERS

EXAMPLE 1 Circle each number that is a multiple of 3.

5, 15, 20, 24, 31, 45, 68, 87.

Solution: If a number is a multiple of 3, then the sum of its digits is divisible by 3. Thus, the numbers are 15, 24, 45, and 87.

EXAMPLE 2 Circle each number that is a multiple of 7.

15, 21, 32, 35, 170, 223, 511, 790, 868

Solution: Test by division. If the remainder is 0, the number is a multiple of 7. Thus, the numbers are 21, 35, 511, and 868.

EXAMPLE 3 Circle each number that is a multiple of 3, 5, and 7.

210, 264, 280, 301, 335, 420, 510

Solution: Only numbers divisible by 5, with final digit 0 or 5, should be tested for divisibility by 3 and by 7. The numbers are 210 and 420.

MULTIPLICATION OF WHOLE NUMBERS

EXAMPLE 1 Use multiplication to complete the chart.

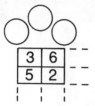

Solution:
$3 \times 6 = 18$	$15 \times 12 = 180$	
$5 \times 2 = 10$	$18 \times 10 = 180$	
$3 \times 5 = 15$	$5 \times 6 = 30$	
$6 \times 2 = 12$	$2 \times 3 = 6$	
	$6 \times 30 = 180$	

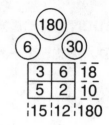

FACTORS OF WHOLE NUMBERS

EXAMPLE 1 List all of the factors of 40.

Solution: Begin with 1. Decide whether each number is a factor of 40.
1 is a factor of 40 since $1 \times 40 = 40$.
2 is a factor of 40 since $2 \times 20 = 40$.
3 is *not* a factor of 40.
4 is a factor of 40 since $4 \times 10 = 40$.
5 is a factor of 40 since $5 \times 8 = 40$.
6 is *not* a factor of 40.
7 is *not* a factor of 40.
Since 7×7 is greater than 40, all of the factors have been found.
They are: 1, 2, 4, 5, 8, 10, 20, and 40.

GREATEST COMMON FACTOR

EXAMPLE 1 Find the greatest number that is a factor of both 40 and 45.

Solution: The factors of 40 are: 1, 2, 4, 5, 8, 10, 20, and 40.
The factors of 45 are: 1, 3, 5, 9, 15, and 45.
The greatest number that is a factor of both 40 and 45 is 5.

MULTIPLICATION AND DIVISION OF WHOLE NUMBERS

EXAMPLE 1 Use multiplication and division to complete the chart.

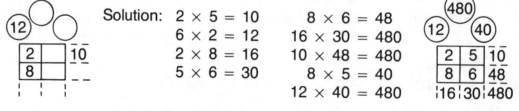

Solution:
$2 \times 5 = 10$	$8 \times 6 = 48$	
$6 \times 2 = 12$	$16 \times 30 = 480$	
$2 \times 8 = 16$	$10 \times 48 = 480$	
$5 \times 6 = 30$	$8 \times 5 = 40$	
	$12 \times 40 = 480$	

FACTORING

EXAMPLE 1 Write 385 as a product of prime numbers.

Solution: $385 \div 5 = 77$ or $385 = 5 \times 77$
$77 \div 7 = 11$ or $385 = 5 \times 7 \times 11$

EXPONENTS

EXAMPLE 1 Complete the following chart.

	Base	Exponent	Is read as	Value
6^2				

Solution: In the expression 6^2, 6 is the base and 2 is the exponent. The exponent tells the number of times the base is used as a factor. 6^2 means 6×6 and is read 6 to the second power. The value of 6^2 is 6×6 or 36.

vi

PRODUCTS OF WHOLE NUMBERS

EXAMPLE 1 Find 7325×5, 47×96, 3216×12, and 604×328.

Solution:

$$
\begin{array}{r}
7325 \\
\times\ 5 \\
\hline
36625
\end{array}
\qquad
\begin{array}{r}
47 \\
\times\ 96 \\
\hline
282 \\
423 \\
\hline
4512
\end{array}
\qquad
\begin{array}{r}
3216 \\
\times\ 12 \\
\hline
6432 \\
3216 \\
\hline
38592
\end{array}
\qquad
\begin{array}{r}
604 \\
\times\ 328 \\
\hline
4832 \\
1208 \\
1812 \\
\hline
198112
\end{array}
$$

FACTORING WHOLE NUMBERS

EXAMPLE 1 Write 496 as a product of prime numbers. Use exponents if you can.

Solution: $496 \div 2 = 248$
$248 \div 2 = 124$
$124 \div 2 = 62$
$62 \div 2 = 31$
$496 = 2 \times 2 \times 2 \times 2 \times 31$ or $2^4 \times 31$

PLACE VALUE

EXAMPLE 1 Write the number eight thousand, four hundred, fifty-two.

Solution: 8452

EXAMPLE 2 Build a number. Write each digit in place. Put in zeros if you need them.

5 in the thousands place,
7 in the tens place,
4 in the ones place

Solution:

thousands	hundreds	tens	ones
5	0	7	4

DIVISION OF WHOLE NUMBERS

EXAMPLE 1 Find $59,229 \div 9$ and $23,733 \div 81$.

Solution:

$$
\begin{array}{r}
6581 \\
9)\overline{59229} \\
54 \\
\hline
52 \\
45 \\
\hline
72 \\
72 \\
\hline
9 \\
9 \\
\hline
0
\end{array}
\qquad
\begin{array}{r}
293 \\
81)\overline{23733} \\
162 \\
\hline
753 \\
729 \\
\hline
243 \\
243 \\
\hline
0
\end{array}
$$

DIVISION PROBLEMS

EXAMPLE 1 If a nurse earns $15,600 per year, what is his monthly salary?

Solution: There are 12 months in a year.

$15,600 \div 12 = 1300

STUDENT RECORD SHEET

Worksheets Completed

Page Number

1	3	5	7		9
2	4	6	8		10
11	13	15	17		19
12	14	16	18		20
21	23	25	27		29
22	24	26	28		30
31	33	35	37		39
32	34	36	38		40
41	43	45	47		49
42	44	46	48		50
51	53	55	57		59
52	54	56	58		60
61	63	65	67		69
62	64	66	68		70
71	73	75	77		79
72	74	76	78		80
81	83	85	87		89
82	84	86	88		90
91	93	95	97		99
92	94	96	98		100

Daily Quiz Grades

No.	Score

Check List Skill Mastered

Date

- ☐ addition
- ☐ subtraction
- ☐ addition and subtraction
- ☐ multiples
- ☐ recognizing multiples
- ☐ multiplication
- ☐ factors
- ☐ greatest common factor
- ☐ multiplication and division
- ☐ factoring into primes
- ☐ exponents
- ☐ products
- ☐ factoring using exponents
- ☐ place value
- ☐ division
- ☐ division problems

Notes

_____

_____

_____

_____

_____

_____

_____

Addition of Whole Numbers

Name _____

Date _____

Add.

1.	2.	3.	4.	5.	6.
3	36	4	243	152	73
2	23	53	167	196	195
5	62	26	592	51	4
7	45	27	+367	267	625
+9	+31	+ 9		+ 92	+ 31
26					

7.
```
         235
      61,729
      43,670
  + 2,160,014
```

8.
```
     367,152
   + 403,709
```

9.
```
         367
      42,103
   + 352,174
```

10.
```
     538,126
      30,081
   + 621,560
```

Complete each of the following.

11. 23 + 16 + 29 = _____

12. 82 + 17 + 16 = _____

13. 35 + 122 + 432 = _____

14. 26 + 3217 + 85 = _____

15. 192 + 533 + 804 = _____

16. 4782 + 8941 = _____

17. Find the total of 8, 13, and 42. _____

18. Find the sum of the following three numbers: 53, 427, and 692. _____

19. Find the total of the following two numbers: 73,249 and 16,127. _____

20. Find the sum of 48, 54, and 163. _____

21. Find the total of the following three numbers: 862, 533, and 926. _____

22. Find the sum of the following two numbers: 93,492 and 22,721. _____

Subtraction of Whole Numbers

Name _____

Date _____

Subtract.

1.   325
    − 63
    ‾‾‾‾‾
      262

2.   840
    −156
    ‾‾‾‾‾

3.   903
    −504
    ‾‾‾‾‾

4.   754
    − 79
    ‾‾‾‾‾

5.   636
    −458
    ‾‾‾‾‾

6.   1097
    − 889
    ‾‾‾‾‾

7.   36,002
    − 2,571
    ‾‾‾‾‾‾‾

8.   56,037
    − 4,912
    ‾‾‾‾‾‾‾

9.   63,394
    − 5,118
    ‾‾‾‾‾‾‾

10. 71,025
    − 3,076
    ‾‾‾‾‾‾‾

Complete each of the following.

11. $307,124 - 2,351 = $ _____

12. $409,672 - 51,433 = $ _____

13. $397,042 - 37,156 = $ _____

14. $660,022 - 53,367 = $ _____

15. Find the difference between 639 and 42. _____

16. 6312 is how much more than 4135? _____

17. Subtract 639 from 1247. _____

18. What number is 17 less than 128? _____

Addition of Whole Numbers

Add.

1.	**2.**	**3.**	**4.**	**5.**	**6.**
8	68	32	642	61	6
7	91	8	438	455	449
9	83	11	141	821	75
4	42	7	120	79	209
+2	+22	+29	+709	+996	+ 87
30					

7. 7,008,635
67,549
538
+ _____ 88,426

8. 801,106
+573,749
_____

9. 807,116
7,562,478
+ _____ 787

10. 1,734,895
496,555
+ 856,432
_____

Complete each of the following.

11. 62 + 89 + 74 = _____

12. 51 + 43 + 36 = _____

13. 27 + 582 + 643 = _____

14. 608 + 144 + 286 = _____

15. 74 + 4136 + 58 = _____

16. 5793 + 9142 = _____

17. Find the total of 7, 12, and 163. _____

18. Find the sum of the following three numbers: 68, 783, and 857. _____

19. Find the total of the following two numbers: 49,642 and 17,762. _____

20. Find the sum of 96, 45, and 841. _____

21. Find the total of the following three numbers: 682, 336, and 127. _____

22. Find the sum of the following two numbers: 89,991 and 47,026. _____

Name _____

Date _____

Subtract.

1. 840	2. 908	3. 703	4. 432	5. 871	6. 506
− 25	−560	−405	− 68	−593	− 99
815					

7. 68,004
  − 3,265

8. 98,750
  − 5,164

9. 54,587
  − 8,196

10. 87,075
  − 7,406

Complete each of the following.

11. 807,124 − 6,421 = _____

13. 664,329 − 541,007 = _____

12. 908,427 − 35,726 = _____

14. 870,705 − 21,438 = _____

15. Find the difference between 172 and 60. _____

16. 3764 is how much more than 2466? _____

17. What number is 42 less than 231? _____

18. Subtract 842 from 1507. _____

Addition of Whole Numbers

Add

1.	2.	3.	4.	5.	6.
1	62	48	608	63	68
7	87	31	911	476	17
9	59	8	426	515	6
8	46	7	+342	85	815
+4	+91	+42		+913	+987
29					

7. 821
   7,142,648
     51,874
   +   62,899

8. 171,743
    +819,467

9. 801,174
     72,403
   +   788

10. 644,577
     8,797,465
    +1,443,216

Complete each of the following.

11. 41 + 27 + 33 = _____

12. 67 + 95 + 26 = _____

13. 223 + 86 + 117 = _____

14. 1778 + 64 + 846 = _____

15. 790 + 175 + 237 = _____

16. 4086 + 5737 = _____

17. Find the total of 7, 27, and 408. _____

18. Find the sum of the following three numbers: 87, 123, and 681. _____

19. Find the total of the following two numbers: 87,108 and 27,849. _____

20. Find the total of the following three numbers: 179, 461, and 848. _____

21. Find the sum of 31, 28, and 101. _____

22. Find the sum of the following two numbers: 18,608 and 39,786. _____

Name _____

Date _____

Subtract.

1. $\begin{array}{r} 196 \\ -\ 39 \\ \hline 157 \end{array}$	**2.** $\begin{array}{r} 723 \\ -106 \\ \hline \end{array}$	**3.** $\begin{array}{r} 508 \\ -219 \\ \hline \end{array}$	**4.** $\begin{array}{r} 874 \\ -\ 63 \\ \hline \end{array}$	**5.** $\begin{array}{r} 779 \\ -598 \\ \hline \end{array}$	**6.** $\begin{array}{r} 1051 \\ -\ 998 \\ \hline \end{array}$

7. $\begin{array}{r} 98,010 \\ -\ 1,417 \\ \hline \end{array}$	**8.** $\begin{array}{r} 97,064 \\ -\ 4,776 \\ \hline \end{array}$	**9.** $\begin{array}{r} 52,311 \\ -\ 4,879 \\ \hline \end{array}$	**10.** $\begin{array}{r} 45,761 \\ -\ 9,832 \\ \hline \end{array}$

Complete each of the following.

11. 480,716 − 3274 = _____

13. 660,398 − 496,237 = _____

12. 689,403 − 87432 = _____

14. 377,895 − 58,876 = _____

15. Find the difference between 820 and 79. _____

16. 8713 is how much more than 5369? _____

17. Subtract 402 from 1304. _____

18. What number is 42 less than 805? _____

Addition of Whole Numbers

Add.

1.	2.	3.	4.	5.	6.
1	63	87	877	61	62
7	37	91	429	373	519
4	45	8	367	842	904
5	87	27	+487	57	2
+6	+21	+ 9		+196	+375
23					

7.
```
    50,073
       242
 2,874,061
+   37,765
```

8.
```
  124,264
 +307,789
```

9.
```
      844
  604,721
+  51,572
```

10.
```
   776,845
 4,153,271
+8,963,422
```

Complete each of the following.

11. 86 + 79 + 23 = _____

12. 98 + 85 + 12 = _____

13. 27 + 201 + 877 = _____

14. 1479 + 29 + 129 = _____

15. 291 + 276 + 642 = _____

16. 4019 + 8765 = _____

17. Find the total of 7, 39, and 998. _____

18. Find the sum of the following three numbers: 533, 77, and 508. _____

19. Find the total of the following two numbers: 83,704 and 27,705. _____

20. Find the sum of 86, 41, and 109. _____

21. Find the total of the following three numbers: 907, 517, and 201. _____

22. Find the sum of the following two numbers: 16,426 and 86,717. _____

Subtraction of Whole Numbers

Name _____

Date _____

Subtract.

1.	826	2.	920	3.	162	4.	951	5.	235	6.	1296
	− 72		−262		−153		− 89		−178		− 778
	754										

7.	64,005	8.	67,808	9.	54,379	10.	47,058
	− 6,257		− 2,723		− 2,189		− 2,779

Complete each of the following.

11. 103,942 − 2,265 = _____

13. 105,527 − 104,219 = _____

12. 842,177 − 32,168 = _____

14. 810,775 − 21,664 = _____

15. Find the difference between 821 and 554. _____

16. 8072 is how much more than 3575? _____

17. Subtract 208 from 1728. _____

18. What number is 29 less than 804? _____

8

Addition of Whole Numbers

Name _____

Date _____

Add.

1.	**2.**	**3.**	**4.**	**5.**	**6.**
3	62	2	646	22	75
7	97	71	271	509	5
5	89	8	993	279	69
9	46	68	+173	25	984
+4	+21	+37		+873	+454
28					

7. 6,589,025
      114
  38,878
+ 46,452

8. 215,310
+919,396

9. 372,828
   984
+ 65,055

10. 8,960,812
2,474,298
+ 872,573

Complete each of the following.

11. 93 + 28 + 97 = _____

12. 86 + 50 + 63 = _____

13. 765 + 133 + 74 = _____

14. 6152 + 99 + 142 = _____

15. 846 + 702 + 198 = _____

16. 1653 + 6774 = _____

17. Find the total of 6, 23, and 49. _____

18. Find the sum of the following three numbers: 610, 57, and 960. _____

19. Find the total of the following two numbers: 86,498 and 51,027. _____

20. Find the sum of 76, 65, and 475. _____

21. Find the total of the following three numbers: 727, 459, and 243. _____

22. Find the sum of the following two numbers: 28,368 and 64,712. _____

Name _____

Date _____

Subtract.

| **1.** 841
 – 12
 829 | **2.** 729
 –399 | **3.** 205
 –118 | **4.** 942
 – 83 | **5.** 962
 –736 | **6.** 1063
 – 978 |

| **7.** 50,851
 – 1,972 | **8.** 44,022
 – 9,153 | **9.** 63,934
 – 8,196 | **10.** 45,176
 – 7,406 |

Complete each of the following.

11. 380,324 – 4,736 = _____   **13.** 322,085 – 97,808 = _____

12. 985,206 – 92,817 = _____   **14.** 513,704 – 275,925 = _____

15. Find the difference between 902 and 58. _____

16. 2553 is how much more than 1175? _____

17. Subtract 539 from 994. _____

18. What number is 12 less than 593? _____

Name _____

Addition and Subtraction of Whole Numbers Date _____

Write the sum and difference for each pair of numbers.

1. 4 and 2 $4 + 2 = 6$ $4 - 2 = 2$	**11.** 7 and 6
2. 9 and 9	**12.** 8 and 3
3. 8 and 4	**13.** 0 and 0
4. 5 and 3	**14.** 9 and 8
5. 7 and 5	**15.** 6 and 3
6. 8 and 6	**16.** 5 and 4
7. 2 and 1	**17.** 7 and 9
8. 8 and 0	**18.** 4 and 4
9. 9 and 1	**19.** 8 and 8
10. 6 and 5	**20.** 9 and 5

Addition and Subtraction of Whole Numbers

Study the examples to learn how to complete the charts using addition and subtraction.

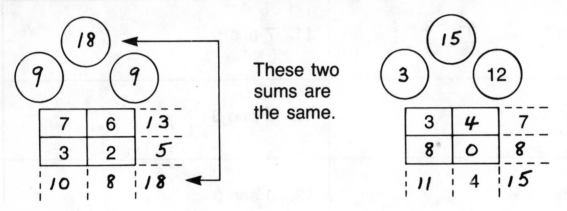

These two sums are the same.

Use addition and subtraction to complete each of the following.

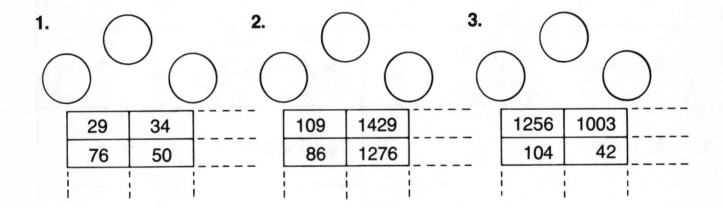

1.

29	34
76	50

2.

109	1429
86	1276

3.

1256	1003
104	42

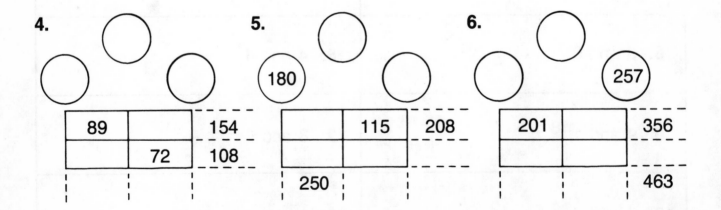

4.

89		154
	72	108

5.

	115	208

250

6.

201		356

463

Addition and Subtraction of Whole Numbers Date _____

Write the sum and difference for each pair of numbers.

1. 13 and 7 $13+7=20$ $13-7=6$	**11.** 14 and 10
2. 16 and 2	**12.** 21 and 1
3. 15 and 12	**13.** 16 and 9
4. 11 and 8	**14.** 17 and 4
5. 20 and 2	**15.** 11 and 7
6. 14 and 1	**16.** 16 and 5
7. 18 and 11	**17.** 20 and 9
8. 14 and 0	**18.** 13 and 8
9. 19 and 18	**19.** 18 and 2
10. 17 and 12	**20.** 10 and 6

Name _____

Addition and Subtraction of Whole Numbers Date _____

Study the examples to learn how to complete the charts using addition and subtraction.

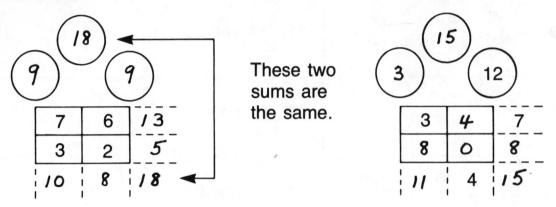

These two sums are the same.

Use addition and subtraction to complete each of the following.

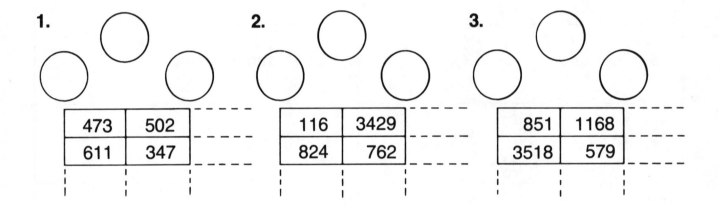

1.

473	502
611	347

2.

116	3429
824	762

3.

851	1168
3518	579

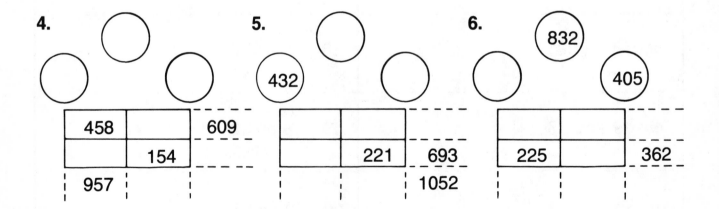

4.

458		609
	154	
957		

5. 432

	221	693
		1052

6. 832 / 405

225		362

Name _____

Addition and Subtraction of Whole Numbers Date _____

Write the sum and difference for each pair of numbers.

1. 50 and 22 $50 + 22 = 72$ $50 - 22 = 28$	**11.** 34 and 26
2. 27 and 23	**12.** 66 and 24
3. 38 and 22	**13.** 28 and 12
4. 40 and 26	**14.** 46 and 24
5. 36 and 24	**15.** 22 and 18
6. 83 and 27	**16.** 26 and 24
7. 50 and 23	**17.** 50 and 27
8. 28 and 22	**18.** 26 and 10
9. 33 and 27	**19.** 60 and 22
10. 40 and 23	**20.** 26 and 14

Addition and Subtraction of Whole Numbers Date _____

Study the examples to learn how to complete the charts using addition and subtraction.

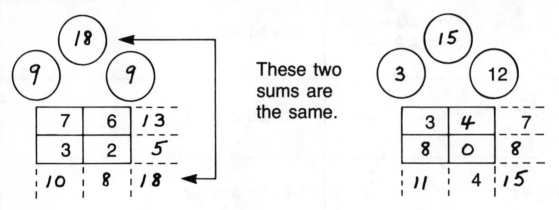

These two sums are the same.

Use addition and subtraction to complete each of the following.

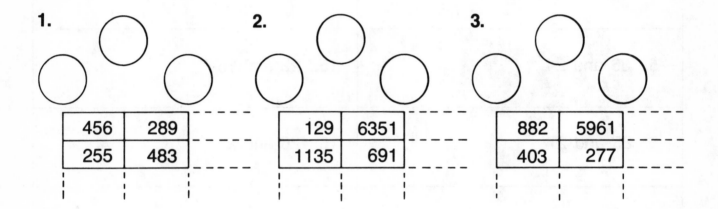

1.

456	289	
255	483	

2.

129	6351	
1135	691	

3.

882	5961	
403	277	

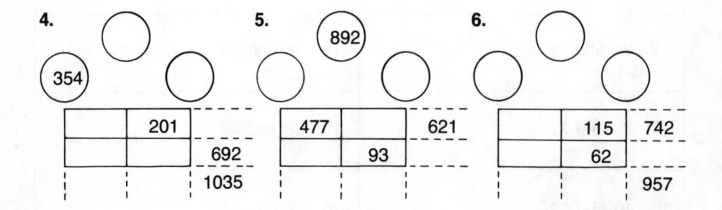

4. 354

	201	
	692	
		1035

5. 892

477		621
	93	

6.

	115	742
	62	
		957

Addition and Subtraction of Whole Numbers Date _____

Write the sum and difference for each pair of numbers.

1. 67 and 23 $67+23=90$ $67-23=44$	**11.** 30 and 26
2. 28 and 20	**12.** 23 and 17
3. 50 and 25	**13.** 80 and 27
4. 51 and 29	**14.** 58 and 22
5. 72 and 28	**15.** 45 and 25
6. 44 and 26	**16.** 93 and 27
7. 48 and 22	**17.** 24 and 20
8. 60 and 28	**18.** 42 and 28
9. 35 and 25	**19.** 82 and 28
10. 50 and 24	**20.** 60 and 27

Addition and Subtraction of Whole Numbers Date _____

Study the examples to learn how to complete the charts using addition and subtraction.

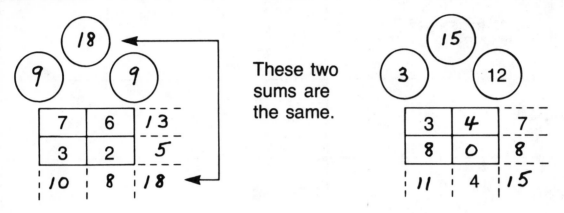

These two sums are the same.

Use addition and subtraction to complete each of the following.

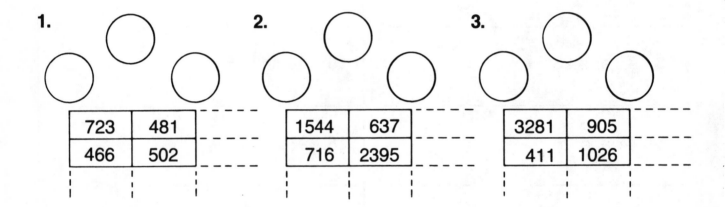

1.

723	481	
466	502	

2.

1544	637	
716	2395	

3.

3281	905	
411	1026	

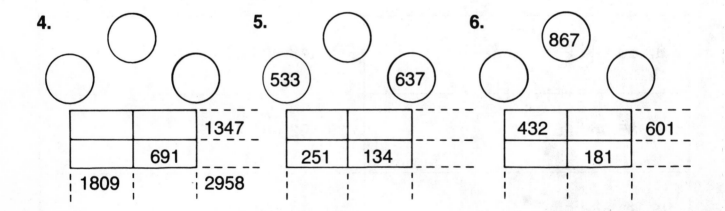

4.

		1347
	691	
1809		2958

5. 533 637

251	134	

6. 867

432		601
	181	

Name _____

Addition and Subtraction of Whole Numbers Date _____

Write the sum and difference for each pair of numbers.

1. 7 and 5	**11.** 19 and 8
2. 4 and 4	**12.** 18 and 7
3. 13 and 7	**13.** 21 and 9
4. 17 and 4	**14.** 55 and 25
5. 20 and 9	**15.** 40 and 26
6. 20 and 11	**16.** 22 and 18
7. 18 and 9	**17.** 44 and 26
8. 21 and 12	**18.** 50 and 25
9. 14 and 4	**19.** 58 and 22
10. 18 and 16	**20.** 82 and 28

Study the examples to learn how to complete the charts using addition and subtraction.

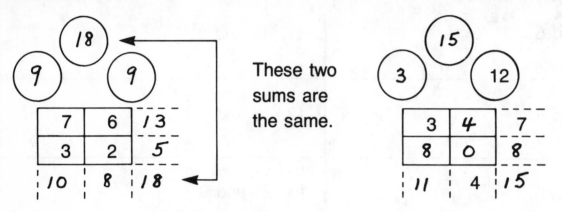

These two sums are the same.

Use addition and subtraction to complete each of the following

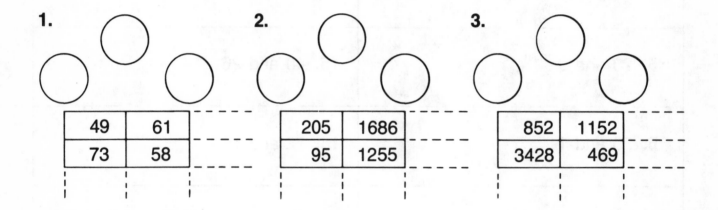

1.

49	61
73	58

2.

205	1686
95	1255

3.

852	1152
3428	469

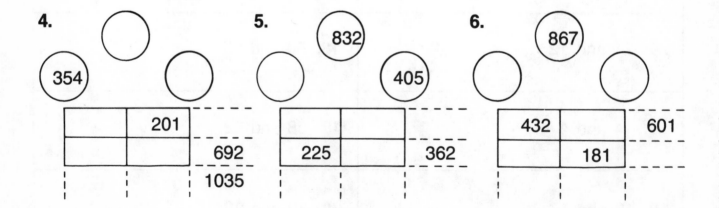

4.

354

	201
	692

1035

5.

832   405

225	362

6.

867

432		601
	181	

Write the facts about multiples of these numbers.

1. 2	*The final digit of the numbers must be 0, 2, 4, 6, or 8.*
2. 3	
3. 5	

Complete each of the following.

4. List 12 numbers between 100 and 200 that are multiples of 2.

___ ___ ___ ___ ___ ___ ___ ___ ___ ___ ___ ___

5. List 12 numbers between 100 and 200 that are multiples of 3.

___ ___ ___ ___ ___ ___ ___ ___ ___ ___ ___ ___

6. List 12 numbers between 100 and 200 that are multiples of 5.

___ ___ ___ ___ ___ ___ ___ ___ ___ ___ ___ ___

Recognizing Multiples of Whole Numbers

Name _____

Date _____

Complete each of the following.

1. Circle each number that is a multiple of 2.

(16)   95   43   71   129   (186)   307

150   296   317   115   426   375   810

2. Circle each number that is a multiple of 3.

72   63   40   87   125   250   314

269   143   385   114   363   825   415

3. Circle each number that is a multiple of 5.

97   36   85   42   125   160   180

295   162   480   325   814   722   365

4. Circle each number that is a multiple of both 2 and 3.

86   90   117   156   148   214   390

476   826   516   386   402   377   415

5. Circle each number that is a multiple of both 2 and 5.

84   125   250   355   160   242   310

145   220   400   312   525   614   380

6. Circle each number that is a multiple of both 3 and 5.

431   565   632   875   436   920   414

630   577   822   702   535   901   300

Multiples of Whole Numbers

Complete each of the following.

---

1. Write 12 numbers between 150 and 225 that are multiples of 2.

152 ___ ___ ___ ___ ___ ___ ___ ___ ___ ___ ___

---

2. Write 12 numbers between 150 and 225 that are multiples of 3.

___ ___ ___ ___ ___ ___ ___ ___ ___ ___ ___ ___

---

3. Write 12 numbers between 150 and 225 that are multiples of 5.

___ ___ ___ ___ ___ ___ ___ ___ ___ ___ ___ ___

---

4. Write 12 numbers between 100 and 200 that are multiples of 7.

___ ___ ___ ___ ___ ___ ___ ___ ___ ___ ___ ___

---

5. Circle each number that is a multiple of 2.

276	431	565	828	623	322	907
675	810	147	832	944	500	637

---

6. Circle each number that is a multiple of 3.

372	417	893	607	456	915	233
497	281	189	506	324	811	702

Recognizing Multiples of Whole Numbers

Complete each of the following.

1. Circle each number that is a multiple of 5.

407	926	(255)	461	(830)	417	700	505
962	714	535	609	960	343		

2. Circle each number that is a multiple of 7.

28	43	60	35	80	49	70	38	67
63	14	78	91	100	107	40	98	

3. Circle each number that is a multiple of both 2 and 3.

182	342	693	127	834	618	413
298	500	604	822	212	561	516

4. Circle each number that is a multiple of both 2 and 5.

160	315	244	562	700	395	812
470	620	861	609	505	417	900

5. Circle each number that is a multiple of both 3 and 5.

255	361	150	210	375	804	345
105	453	394	580	507	810	434

6. Circle each number that is a multiple of both 2 and 7.

42	81	95	44	83	28	19	54	
82	66	77	88	49	98	62	100	105

7. Circle each number that is a multiple of both 5 and 7.

85	92	35	102	77	70	86	105
114	122	125	130	58	69	50	

24

Name _____

Date _____

Complete each of the following.

---

1. List 12 numbers between 200 and 250 that are multiples of 2.

_202_ ___ ___ ___ ___ ___ ___ ___ ___ ___ ___ ___

---

2. List 12 numbers between 200 and 250 that are multiples of 3.

___ ___ ___ ___ ___ ___ ___ ___ ___ ___ ___ ___

---

3. List 12 numbers between 200 and 300 that are multiples of 5.

___ ___ ___ ___ ___ ___ ___ ___ ___ ___ ___ ___

---

4. List 12 numbers between 100 and 200 that are multiples of 4.

___ ___ ___ ___ ___ ___ ___ ___ ___ ___ ___ ___

---

5. How can you recognize numbers greater than 100 that are multiples

of 4? _____

---

6. List 12 numbers between 200 and 300 that are multiples of 7.

___ ___ ___ ___ ___ ___ ___ ___ ___ ___ ___ ___

Name _____

Date _____

Complete each of the following.

1. Circle the numbers that are multiples of both 2 and 3.

251   (324)   621   (234   327   813   618

432   676   955   345   701   900   562

2. Circle the numbers that are multiples of both 2 and 7.

32   42   55   67   70   35   77   84   96

14   95   98   101   123   63   72   135

3. Circle the numbers that are multiples of both 3 and 7.

56   42   97   84   36   91   28   35   77

86   99   101   126   55   63   90   105

4. Circle the numbers that are multiples of both 5 and 7.

135   70   114   135   105   83   210

107   82   125   150   217   304   280   140

5. Circle the numbers that are multiples of 4.

114   126   132   146   156   181   150

175   116   111   143   180   112   155

6. Numbers that are multiples of both 2 and 3 are also multiples of _____.

7. Numbers that are multiples of both 2 and 7 are also multiples of _____.

8. Numbers that are multiples of both 2 and 5 are also multiples of _____.

9. Numbers that are multiples of both 3 and 5 are also multiples of _____.

10. Numbers that are multiples of both 3 and 7 are also multiples of _____.

Complete each of the following.

| 25 | 38 | 45 | 62 | 84 | 30 | 73 | 39 | 54 | 91 | 63 | 32 | 55 | 94 |

1. List the numbers from above that are multiples of 2.

_38_ _____

2. List the numbers from above that are multiples of 3.

_____

3. List the numbers from above that are multiples of 5.

_____

4. List the numbers from above that are multiples of 7.

_____

Recognizing Multiples of Whole Numbers

Name _____

Date _____

Complete each of the following.

| 70 | 195 | 168 | 182 | 330 | 154 | 162 | 234 | 183 | 401 | 323 | 304 |

1. List the numbers from above that are multiples of 2.

    70 _____

2. List the numbers from above that are multiples of 3.

    _____

3. List the numbers from above that are multiples of 5.

    _____

4. List the numbers from above that are multiples of 7.

    _____

5. List 12 multiples of 4 between 200 and 300.

    ___ ___ ___ ___ ___ ___ ___ ___ ___ ___ ___ ___

6. Circle the numbers below that are multiples of 4.

    | 226 | 243 | 216 | 285 | 233 | 246 | 214 | 228 | 236 | 217 |
    | 292 | 282 | 284 | 296 | 298 | 202 | 250 | 264 | 229 | 272 |
    | 220 | 267 | 290 | 300 | 301 | 383 | 402 | 361 | 434 | 903 |
    | 506 | 312 | 532 | 716 | 823 | 436 | 535 | 200 | 840 | |

28

Multiples of Whole Numbers

Complete each of the following.

25	38	45	62	84	126	30	76	142	96	130	156
143	70	195	182	330	154	162	235	180	250		
210	200	196	304	327	434	357	343				

1. List the numbers from above that are multiples of 2.

_38_ _____

2. List the numbers from above that are multiples of 3.

_____

3. List the numbers from above that are multiples of 4.

_____

4. List the numbers from above that are multiples of 5.

_____

5. List the numbers from above that are multiples of 7.

_____

6. List the numbers from above that are multiples of both 2 and 3.

_____

Name _____

Date _____

Complete each of the following.

25	38	45	62	84	126	30	76	142	96	130	156
143	70	195	182	330	154	162	235	180	250		
210	200	196	304	327	434	357	343				

1. List the numbers from above that are multiples of both 2 and 5.

_30_ _____

2. List the numbers from above that are multiples of both 2 and 7.

_____

3. List the numbers from above that are multiples of both 3 and 5.

_____

4. List the numbers from above that are multiples of both 3 and 7.

_____

5. List the numbers from above that are multiples of both 5 and 7.

_____

6. List the numbers from above that are multiples of both 4 and 5.

_____

Multiplication of Whole Numbers

Study the example
to learn how to
complete the charts
using multiplication.

2	5	10
8	6	48
16	30	480

Multiply to complete each of the following.

1.

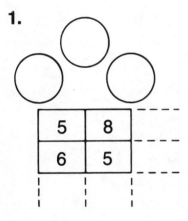

5	8
6	5

2.

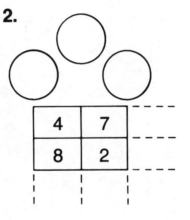

4	7
8	2

3.

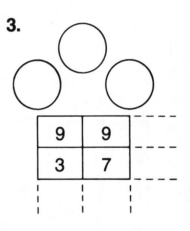

9	9
3	7

4.

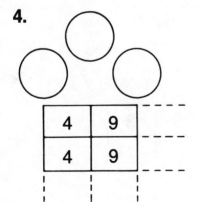

4	9
4	9

5.

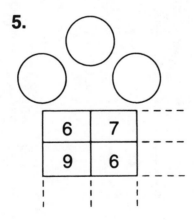

6	7
9	6

6.

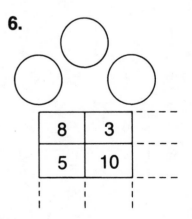

8	3
5	10

31

Name _____

Date _____

Study the example
to learn how to
complete the charts
using multiplication.

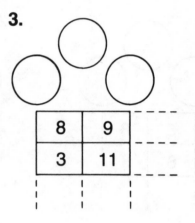

2	5	10
8	6	48
16	30	480

Multiply to complete each of the following.

1.

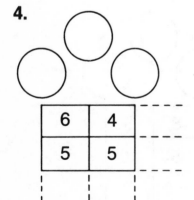

6	2	
4	7	

2.

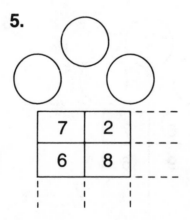

7	5	
8	10	

3.

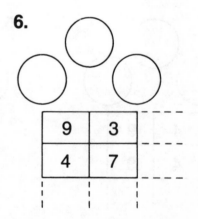

8	9	
3	11	

4.

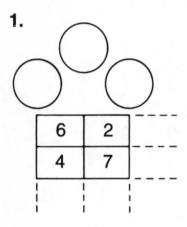

6	4	
5	5	

5.

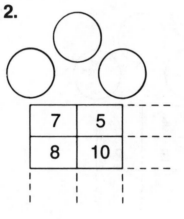

7	2	
6	8	

6.

9	3	
4	7	

Multiplication of Whole Numbers

Name _____

Date _____

Study the example
to learn how to
complete the charts
using multiplication.

```
        (480)
  (12)        (40)
```

2	5	10
8	6	48
16	30	480

Multiply to complete each of the following.

1.

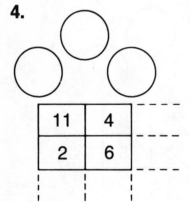

2.

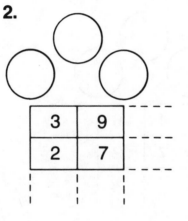

3.

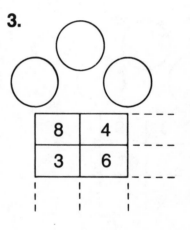

4.

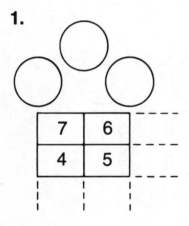

5.

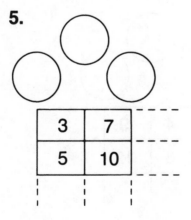

6.

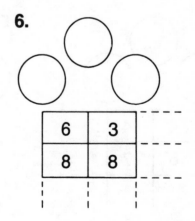

33

Name _____

Date _____

Study the example to learn how to complete the charts using multiplication.

Multiply to complete each of the following.

1.

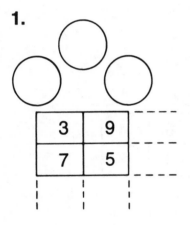

2.

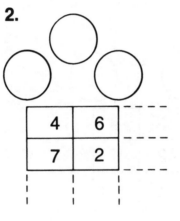

3.

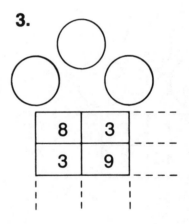

4.

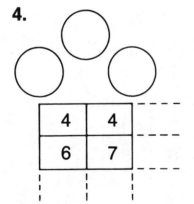

5.

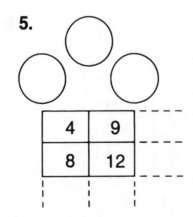

6.

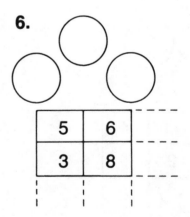

34

Multiplication of Whole Numbers

Study the example
to learn how to
complete the charts
using multiplication.

```
        (480)
   (12)      (40)
```

2	5	10
8	6	48
16	30	480

Multiply to complete each of the following.

1.
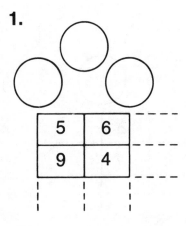

5	6
9	4

2.
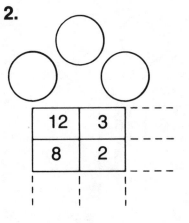

12	3
8	2

3.
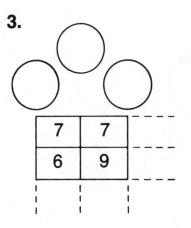

7	7
6	9

4.
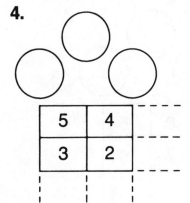

5	4
3	2

5.
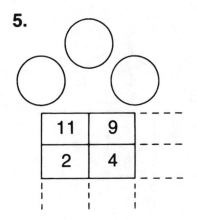

11	9
2	4

6.
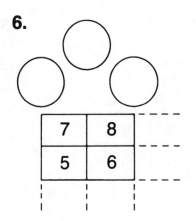

7	8
5	6

Name _____

Date _____

Study the example
to learn how to
complete the charts
using multiplication.

	480	
12		40

2	5	10
8	6	48
16	30	480

Multiply to complete each of the following.

1.

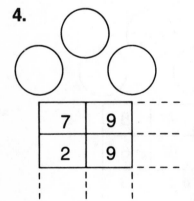

10	4
3	2

2.

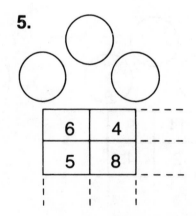

9	6
7	5

3.

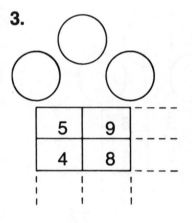

5	9
4	8

4.

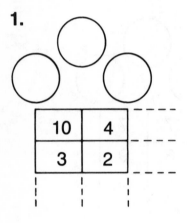

7	9
2	9

5.

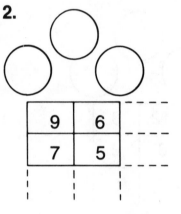

6	4
5	8

6.

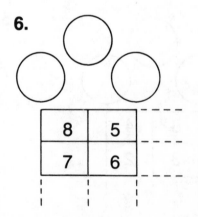

8	5
7	6

Multiplication of Whole Numbers

Name _____

Date _____

Study the example
to learn how to
complete the charts
using multiplication.

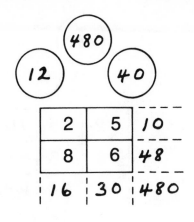

Multiply to complete each of the following.

1.

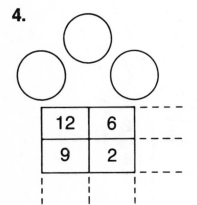

2.

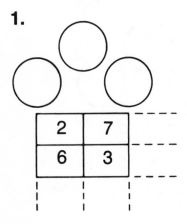

3.

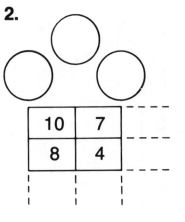

4.

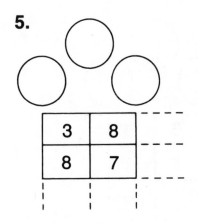

5.

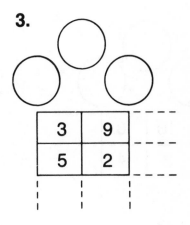

6.

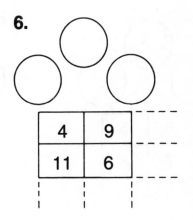

Multiplication of Whole Numbers

Name _____

Date _____

Study the example
to learn how to
complete the charts
using multiplication.

| | 480 | |
| 12 | | 40 |

2	5	10
8	6	48
16	30	480

Multiply to complete each of the following.

1.
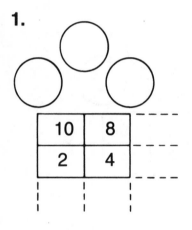

10	8
2	4

2.
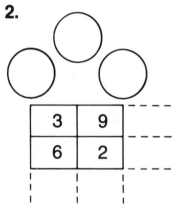

3	9
6	2

3.
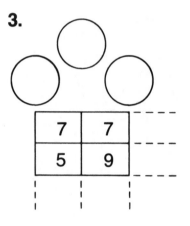

7	7
5	9

4.
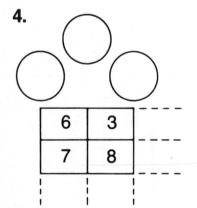

6	3
7	8

5.
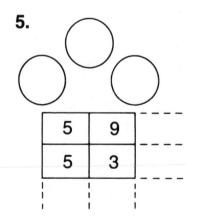

5	9
5	3

6.
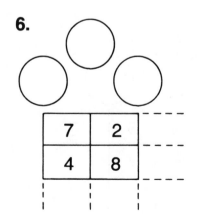

7	2
4	8

Multiplication of Whole Numbers

Name _____

Date _____

Study the example
to learn how to
complete the charts
using multiplication.

| | 480 | |
| 12 | | 40 |

2	5	10
8	6	48
16	30	480

Multiply to complete each of the following.

1.

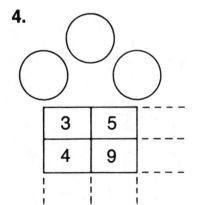

8	3
2	7

2.

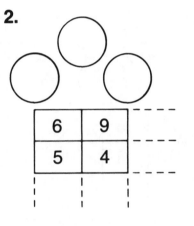

6	9
5	4

3.

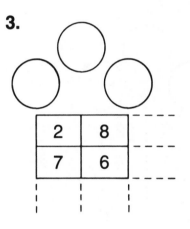

2	8
7	6

4.

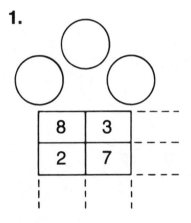

3	5
4	9

5.

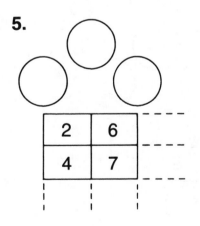

2	6
4	7

6.

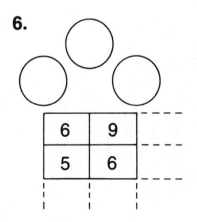

6	9
5	6

Multiplication of Whole Numbers

Name _____

Date _____

Study the example
to learn how to
complete the charts
using multiplication.

```
        (480)
  (12)        (40)
   ┌─────┬─────┐
   │  2  │  5  │ 10
   ├─────┼─────┤
   │  8  │  6  │ 48
   └─────┴─────┘
     16   30   480
```

Multiply to complete each of the following.

1.

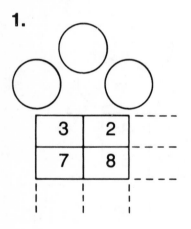

2.

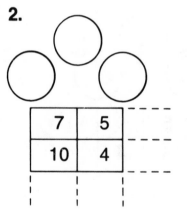

3.

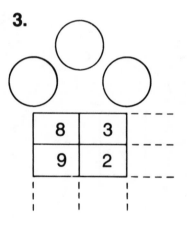

4.

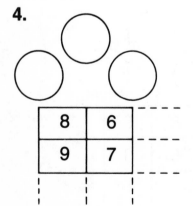

5.

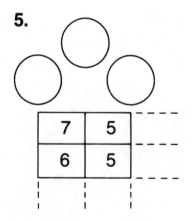

6.

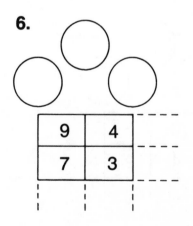

Factors of Whole Numbers

Complete each of the following.

1. factors of 14

$14 = 1 \times \underline{14}$ $\underline{1}$ $\underline{14}$

$14 = 2 \times \underline{7}$ $\underline{2}$ $\underline{7}$

5. factors of 35

$35 = 1 \times \underline{\hspace{1cm}}$ $\underline{\hspace{1cm}}$ $\underline{\hspace{1cm}}$

$35 = 5 \times \underline{\hspace{1cm}}$ $\underline{\hspace{1cm}}$ $\underline{\hspace{1cm}}$

2. factors of 27

$27 = 1 \times \underline{\hspace{1cm}}$ $\underline{\hspace{1cm}}$ $\underline{\hspace{1cm}}$

$27 = 3 \times \underline{\hspace{1cm}}$ $\underline{\hspace{1cm}}$ $\underline{\hspace{1cm}}$

6. factors of 22

$22 = 1 \times \underline{\hspace{1cm}}$ $\underline{\hspace{1cm}}$ $\underline{\hspace{1cm}}$

$22 = 2 \times \underline{\hspace{1cm}}$ $\underline{\hspace{1cm}}$ $\underline{\hspace{1cm}}$

3. factors of 45

$45 = 1 \times \underline{\hspace{1cm}}$ $\underline{\hspace{1cm}}$ $\underline{\hspace{1cm}}$

$45 = 3 \times \underline{\hspace{1cm}}$ $\underline{\hspace{1cm}}$ $\underline{\hspace{1cm}}$

$45 = 5 \times \underline{\hspace{1cm}}$ $\underline{\hspace{1cm}}$ $\underline{\hspace{1cm}}$

7. factors of 16

$16 = 1 \times \underline{\hspace{1cm}}$ $\underline{\hspace{1cm}}$ $\underline{\hspace{1cm}}$

$16 = 2 \times \underline{\hspace{1cm}}$ $\underline{\hspace{1cm}}$ $\underline{\hspace{1cm}}$

$16 = 4 \times \underline{\hspace{1cm}}$ $\underline{\hspace{1cm}}$ $\underline{\hspace{1cm}}$

4. factors of 24

$24 = 1 \times \underline{\hspace{1cm}}$ $\underline{\hspace{1cm}}$ $\underline{\hspace{1cm}}$

$24 = 2 \times \underline{\hspace{1cm}}$ $\underline{\hspace{1cm}}$ $\underline{\hspace{1cm}}$

$24 = 3 \times \underline{\hspace{1cm}}$ $\underline{\hspace{1cm}}$ $\underline{\hspace{1cm}}$

$24 = 4 \times \underline{\hspace{1cm}}$ $\underline{\hspace{1cm}}$ $\underline{\hspace{1cm}}$

8. factors of 30

$30 = 1 \times \underline{\hspace{1cm}}$ $\underline{\hspace{1cm}}$ $\underline{\hspace{1cm}}$

$30 = 2 \times \underline{\hspace{1cm}}$ $\underline{\hspace{1cm}}$ $\underline{\hspace{1cm}}$

$30 = 3 \times \underline{\hspace{1cm}}$ $\underline{\hspace{1cm}}$ $\underline{\hspace{1cm}}$

$30 = 5 \times \underline{\hspace{1cm}}$ $\underline{\hspace{1cm}}$ $\underline{\hspace{1cm}}$

Write the greatest number that is a factor of both numbers.

9. 14 and 45 _1_

10. 24 and 16 _____

11. 30 and 16 _____

12. 45 and 22 _____

13. 16 and 27 _____

14. 35 and 22 _____

15. 45 and 27 _____

16. 30 and 45 _____

Name _____

Date _____

The handwriting shows how the figure at the right is based on multiplication.

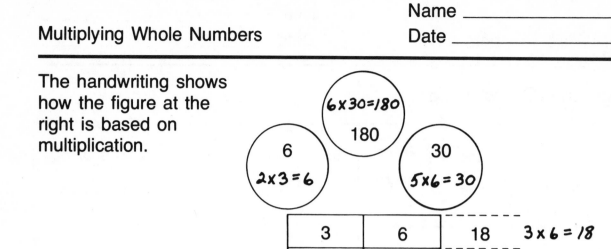

6×30=180
180
6   2×3=6
30   5×6=30

3	6	18	3×6=18
5	2	10	5×2=10
15	12	180	18×10=180

3×5=15   6×2=12   15×12=180

Complete each of the following.

1.

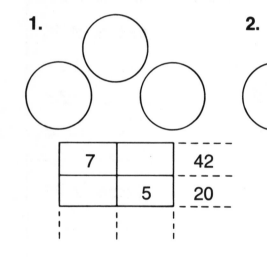

| 7 | | 42 |
| | 5 | 20 |

2.

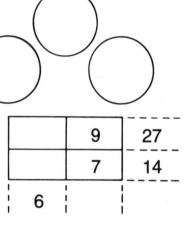

| | 9 | 27 |
| | 7 | 14 |

6

3.

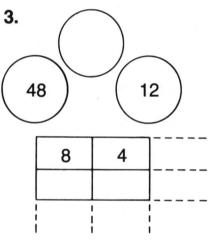

48   12

| 8 | 4 | |

4.

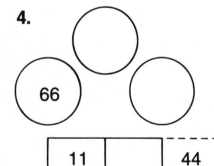

66

| 11 | | 44 |
| 2 | | |

5.

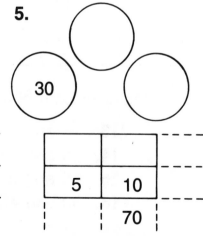

30

| | | |
| 5 | 10 | |

70

6.

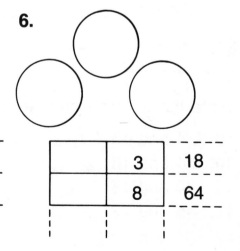

| | 3 | 18 |
| | 8 | 64 |

Factors of Whole Numbers

Name _____

Date _____

Complete each of the following.

1. factors of 25

$25 = 1 \times \underline{25}$ $\underline{1}$ $\underline{25}$

$25 = 5 \times \underline{5}$ $\underline{5}$

5. factors of 39

$39 = 1 \times \underline{\hspace{1.5em}}$ $\underline{\hspace{1.5em}}$ $\underline{\hspace{1.5em}}$

$39 = 3 \times \underline{\hspace{1.5em}}$ $\underline{\hspace{1.5em}}$ $\underline{\hspace{1.5em}}$

2. factors of 46

$46 = 1 \times \underline{\hspace{1.5em}}$ $\underline{\hspace{1.5em}}$ $\underline{\hspace{1.5em}}$

$46 = 2 \times \underline{\hspace{1.5em}}$ $\underline{\hspace{1.5em}}$ $\underline{\hspace{1.5em}}$

6. factors of 62

$62 = 1 \times \underline{\hspace{1.5em}}$ $\underline{\hspace{1.5em}}$ $\underline{\hspace{1.5em}}$

$62 = 2 \times \underline{\hspace{1.5em}}$ $\underline{\hspace{1.5em}}$ $\underline{\hspace{1.5em}}$

3. factors of 32

$32 = 1 \times \underline{\hspace{1.5em}}$ $\underline{\hspace{1.5em}}$ $\underline{\hspace{1.5em}}$

$32 = 2 \times \underline{\hspace{1.5em}}$ $\underline{\hspace{1.5em}}$ $\underline{\hspace{1.5em}}$

$32 = 4 \times \underline{\hspace{1.5em}}$ $\underline{\hspace{1.5em}}$ $\underline{\hspace{1.5em}}$

7. factors of 44

$44 = 1 \times \underline{\hspace{1.5em}}$ $\underline{\hspace{1.5em}}$ $\underline{\hspace{1.5em}}$

$44 = 2 \times \underline{\hspace{1.5em}}$ $\underline{\hspace{1.5em}}$ $\underline{\hspace{1.5em}}$

$44 = 4 \times \underline{\hspace{1.5em}}$ $\underline{\hspace{1.5em}}$ $\underline{\hspace{1.5em}}$

4. factors of 36

$36 = 1 \times \underline{\hspace{1.5em}}$ $\underline{\hspace{1.5em}}$ $\underline{\hspace{1.5em}}$

$36 = 2 \times \underline{\hspace{1.5em}}$ $\underline{\hspace{1.5em}}$ $\underline{\hspace{1.5em}}$

$36 = 3 \times \underline{\hspace{1.5em}}$ $\underline{\hspace{1.5em}}$ $\underline{\hspace{1.5em}}$

$36 = 4 \times \underline{\hspace{1.5em}}$ $\underline{\hspace{1.5em}}$ $\underline{\hspace{1.5em}}$

$36 = 6 \times \underline{\hspace{1.5em}}$ $\underline{\hspace{1.5em}}$

8. factors of 54

$54 = 1 \times \underline{\hspace{1.5em}}$ $\underline{\hspace{1.5em}}$ $\underline{\hspace{1.5em}}$

$54 = 2 \times \underline{\hspace{1.5em}}$ $\underline{\hspace{1.5em}}$ $\underline{\hspace{1.5em}}$

$54 = 3 \times \underline{\hspace{1.5em}}$ $\underline{\hspace{1.5em}}$ $\underline{\hspace{1.5em}}$

$54 = 6 \times \underline{\hspace{1.5em}}$ $\underline{\hspace{1.5em}}$ $\underline{\hspace{1.5em}}$

Write the greatest number that is a factor of both numbers.

9. 32 and 36 $\underline{4}$

10. 36 and 44 $\underline{\hspace{1.5em}}$

11. 39 and 54 $\underline{\hspace{1.5em}}$

12. 62 and 32 $\underline{\hspace{1.5em}}$

13. 46 and 54 $\underline{\hspace{1.5em}}$

14. 36 and 54 $\underline{\hspace{1.5em}}$

15. 25 and 44 $\underline{\hspace{1.5em}}$

16. 32 and 39 $\underline{\hspace{1.5em}}$

Name _____

Date _____

The handwriting shows how the figure at the right is based on multiplication.

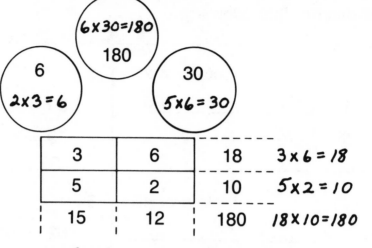

6x30=180
180
6
2x3=6
30
5x6=30

3	6	18	3x6=18
5	2	10	5x2=10
15	12	180	18x10=180

3x5=15 6x2=12 15x12=180

Complete each of the following.

1.

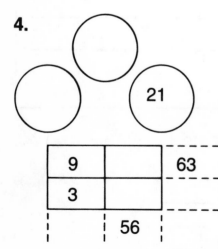

15

| 3 | | 27 |
| 7 | | |

2.

8   42

| 4 | 6 | |
| 7 | | |

3.

8		24
	9	
24		

4.

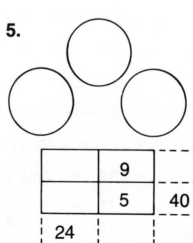

21

9		63
3		
	56	

5.

9   5   40

	9	
	5	40
24		

6.

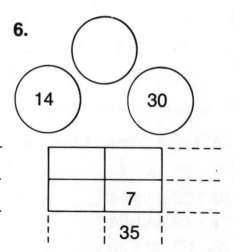

14   30

	7	
	35	

44

Factors of Whole Numbers

Name _____

Date _____

Complete each of the following.

1.	factors of 15
15 = 1 × _15_ _1_ _15_	
15 = 3 × _5_ _3_ _5_	

5.	factors of 26
26 = 1 × ____ ____ ____	
26 = 2 × ____ ____ ____	

2.	factors of 6
6 = 1 × ____ ____ ____	
6 = 2 × ____ ____ ____	

6.	factors of 13
13 = 1 × ____ ____ ____	

3.	factors of 12
12 = 1 × ____ ____ ____	
12 = 2 × ____ ____ ____	
12 = 3 × ____ ____ ____	

7.	factors of 18
18 = 1 × ____ ____ ____	
18 = 2 × ____ ____ ____	
18 = 3 × ____ ____ ____	

4.	factors of 40
40 = 1 × ____ ____ ____	
40 = 2 × ____ ____ ____	
40 = 4 × ____ ____ ____	
40 = 5 × ____ ____ ____	

8.	factors of 42
42 = 1 × ____ ____ ____	
42 = 2 × ____ ____ ____	
42 = 3 × ____ ____ ____	
42 = 6 × ____ ____ ____	

Write the greatest number that is a factor of both numbers.

9. 15 and 12 __3__

10. 18 and 26 ____

11. 13 and 15 ____

12. 40 and 15 ____

13. 6 and 18 ____

14. 13 and 26 ____

15. 42 and 18 ____

16. 6 and 40 ____

Name _____

Date _____

The handwriting shows
how the figure on the
right is based on
multiplication.

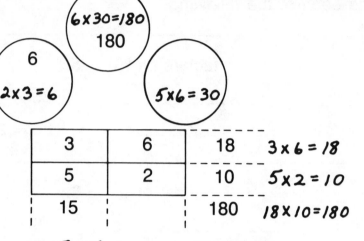

6×30=180
180

6

2×3=6

5×6=30

3	6	18	3×6=18
5	2	10	5×2=10
15		180	18×10=180

3×5=15   6×2=12   15×12=180

Complete each of the following.

1.

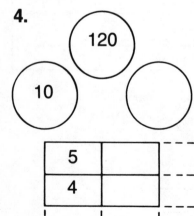

	6	30
45		1080

2.

576

24

12	3	

3.

63

42

7	7	

4.

120

10

5	
4	

5.

44

18

	4	
22	36	

6.

35

48

7	8	

Factors of Whole Numbers

Name _____

Date _____

Complete each of the following.

1.   factors of 35

$35 = 1 \times$ _35_   _1_   _35_

$35 = 5 \times$ _7_   _5_   _7_

5.   factors of 49

$49 = 1 \times$ _____   _____ _____

$49 = 7 \times$ _____   _____

2.   factors of 8

$8 = 1 \times$ _____   _____ _____

$8 = 2 \times$ _____   _____ _____

6.   factors of 21

$21 = 1 \times$ _____   _____ _____

$21 = 3 \times$ _____   _____ _____

3.   factors of 28

$28 = 1 \times$ _____   _____ _____

$28 = 2 \times$ _____   _____ _____

$28 = 4 \times$ _____   _____ _____

7.   factors of 20

$20 = 1 \times$ _____   _____ _____

$20 = 2 \times$ _____   _____ _____

$20 = 4 \times$ _____   _____ _____

4.   factors of 48

$48 = 1 \times$ _____   _____ _____

$48 = 2 \times$ _____   _____ _____

$48 = 3 \times$ _____   _____ _____

$48 = 4 \times$ _____   _____ _____

$48 = 6 \times$ _____   _____ _____

8.   factors of 42

$42 = 1 \times$ _____   _____ _____

$42 = 2 \times$ _____   _____ _____

$42 = 3 \times$ _____   _____ _____

$42 = 6 \times$ _____   _____ _____

Write the greatest number that is a factor of both numbers.

9. 21 and 28 _7_

10. 42 and 21 _____

11. 20 and 48 _____

12. 28 and 49 _____

13. 8 and 28 _____

14. 35 and 42 _____

15. 20 and 35 _____

16. 42 and 8 _____

Name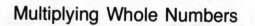

Date _____

The handwriting shows how the figure at the right is based on multiplication.

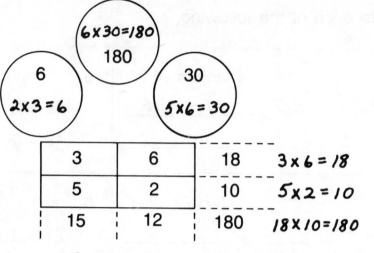

6×30=180
180

6
2×3=6

30
5×6=30

3	6	18	3×6=18
5	2	10	5×2=10
15	12	180	18×10=180

3×5=15 6×2=12 15×12=180

Complete each of the following.

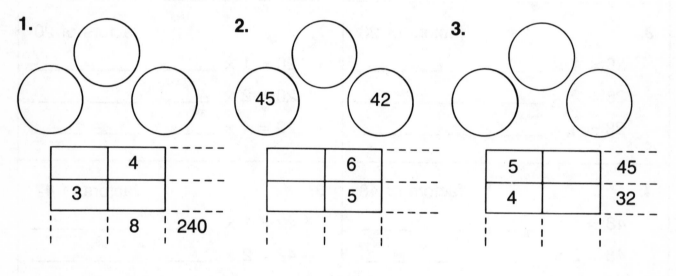

1.

	4	
3		
	8	240

2.

45   42

| | 6 | |
| | 5 | |

3.

| 5 | | 45 |
| 4 | | 32 |

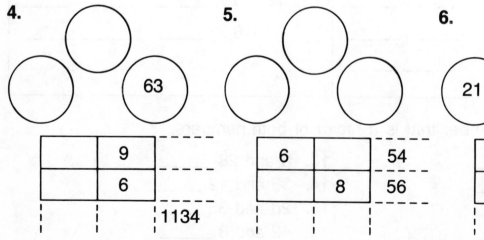

4.

63

	9	
	6	
		1134

5.

| 6 | | 54 |
| | 8 | 56 |

6.

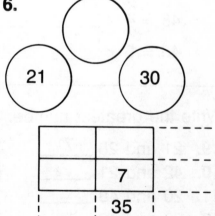

21   30

	7	
	35	

Factors of Whole Numbers

Name _____

Date _____

Complete each of the following.

1. factors of 23 23 = 1 × _23_ _1_ _23_	**5.** factors of 46 46 = 1 × ____ ____ ____ 46 = 2 × ____ ____ ____
2. factors of 14 14 = 1 × ____ ____ ____ 14 = 2 × ____ ____ ____	**6.** factors of 8 8 = 1 × ____ ____ ____ 8 = 2 × ____ ____ ____
3. factors of 50 50 = 1 × ____ ____ ____ 50 = 2 × ____ ____ ____ 50 = 5 × ____ ____ ____	**7.** factors of 52 52 = 1 × ____ ____ ____ 52 = 2 × ____ ____ ____ 52 = 4 × ____ ____ ____
4. factors of 56 56 = 1 × ____ ____ ____ 56 = 2 × ____ ____ ____ 56 = 4 × ____ ____ ____ 56 = 7 × ____ ____ ____	**8.** factors of 75 75 = 1 × ____ ____ ____ 75 = 3 × ____ ____ ____ 75 = 5 × ____ ____ ____

Write the greatest number that is a factor of both numbers.

9. 52 and 56 _4_
10. 75 and 50 ____
11. 56 and 46 ____
12. 8 and 14 ____

13. 23 and 46 ____
14. 50 and 56 ____
15. 14 and 75 ____
16. 56 and 14 ____

49

Multiplying Whole Numbers

Name _____
Date _____

The handwriting shows
how the figure at the
right is based on
multiplication.

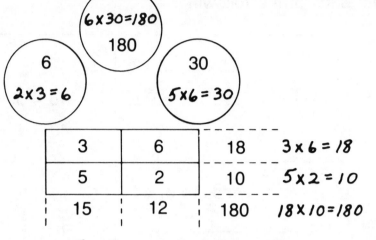

6x30=180
180

6
2x3=6

30
5x6=30

3	6	18	3 x 6 = 18
5	2	10	5 x 2 = 10
15	12	180	18 x 10 = 180

3x5=15   6x2=12   15x12=180

Complete each of the following.

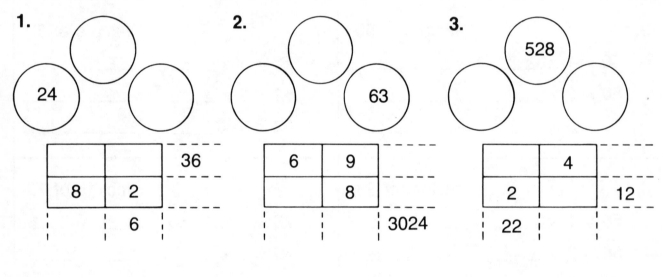

1.

24

		36
8	2	
	6	

2.

63

6	9	
	8	
		3024

3.

528

	4	
2		12
22		

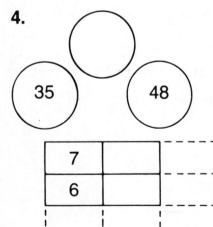
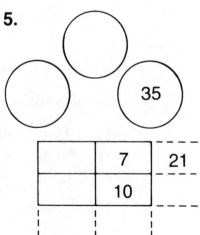
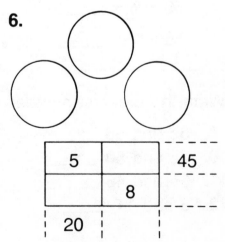

4.

35   48

7		
6		

5.

35

	7	21
	10	

6.

5		45
	8	
20		

50

Name _____

Date _____

The factors of 99 are 1, 3, 9, 11, 33, and 99.
List all the factors for the following numbers.

1. 1 ___/_____

2. 2 _____

3. 3 _____

4. 4 _____

5. 5 _____

6. 6 _____

7. 7 _____

8. 8 _____

9. 9 _____

10. 10 _____

11. 11 _____

12. 12 _____

13. 13 _____

14. 14 _____

15. 15 _____

16. 16 _____

17. 17 _____

18. 18 _____

19. 19 _____

20. 20 _____

21. 21 _____

22. 22 _____

23. 23 _____

24. 24 _____

25. 25 _____

26. 26 _____

27. 27 _____

28. 28 _____

29. 29 _____

30. 30 _____

31. 31 _____

32. 32 _____

Name _____

Date _____

The handwriting shows how the figure at the right is based on multiplication.

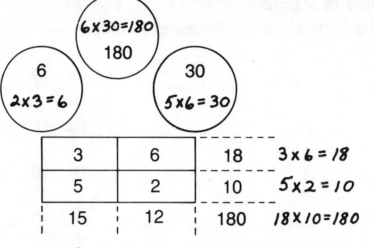

3	6	18	$3 \times 6 = 18$
5	2	10	$5 \times 2 = 10$
15	12	180	$18 \times 10 = 180$

$3 \times 5 = 15 \quad 6 \times 2 = 12 \quad 15 \times 12 = 180$

Complete each of the following.

1.

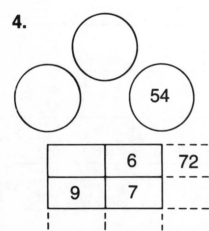

2.

3.

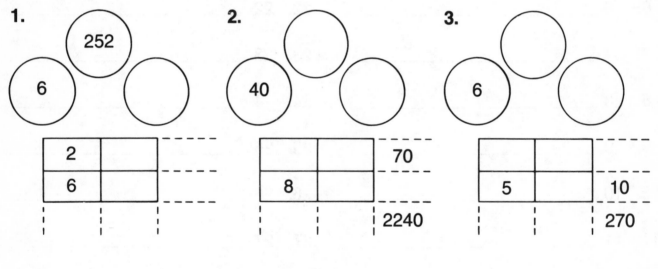

4.

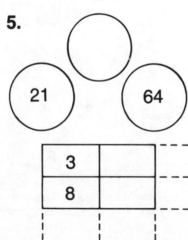

5.

6.

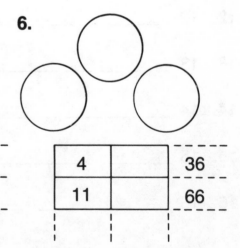

Name _____

Date _____

The factors of 99 are 1, 3, 9, 11, 33, and 99.
List all the factors for the following numbers.

1. 33 ___*1, 3, 11, 33*___

2. 34 _____

3. 35 _____

4. 36 _____

5. 37 _____

6. 38 _____

7. 39 _____

8. 40 _____

9. 41 _____

10. 42 _____

11. 43 _____

12. 44 _____

13. 45 _____

14. 46 _____

15. 47 _____

16. 48 _____

17. 49 _____

18. 50 _____

19. 51 _____

20. 52 _____

21. 53 _____

22. 54 _____

23. 55 _____

24. 56 _____

25. 57 _____

26. 58 _____

27. 59 _____

28. 60 _____

29. 61 _____

30. 62 _____

31. 63 _____

32. 64 _____

Name _____

Date _____

The handwriting shows how the figure at the right is based on multiplication.

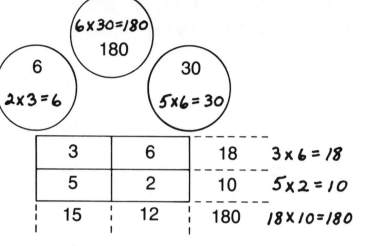

6×30=180
180

6
2×3=6

30
5×6=30

3	6	18	3×6=18
5	2	10	5×2=10
15	12	180	18×10=180

3×5=15 6×2=12 15×12=180

Complete each of the following.

1.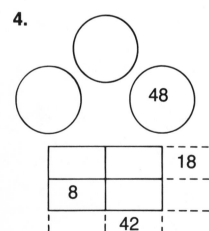

640

40

2		
20		

2.

324

6

3		27

3.

63   35

5	9	

4.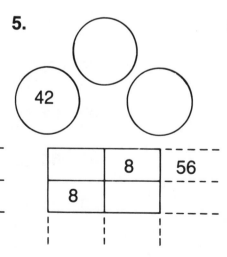

48

		18
8		
	42	

5.

42

	8	56
8		

6.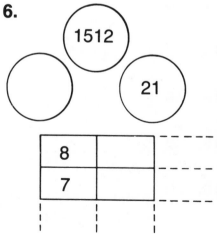

1512

21

8		
7		

The factors of 99 are 1, 3, 9, 11, 33, and 99.
List all the factors for the following numbers.

1. 65 _____ *1, 5, 13, 65* _____

2. 66 _____

3. 67 _____

4. 68 _____

5. 69 _____

6. 70 _____

7. 71 _____

8. 72 _____

9. 73 _____

10. 74 _____

11. 75 _____

12. 76 _____

13. 77 _____

14. 78 _____

15. 79 _____

16. 80 _____

17. 81 _____

18. 82 _____

19. 83 _____

20. 84 _____

21. 85 _____

22. 86 _____

23. 87 _____

24. 88 _____

25. 89 _____

26. 90 _____

27. 91 _____

28. 92 _____

29. 93 _____

30. 94 _____

31. 95 _____

32. 96 _____

Name _____

Date _____

The handwriting shows
how the figure at the
right is based on
multiplication.

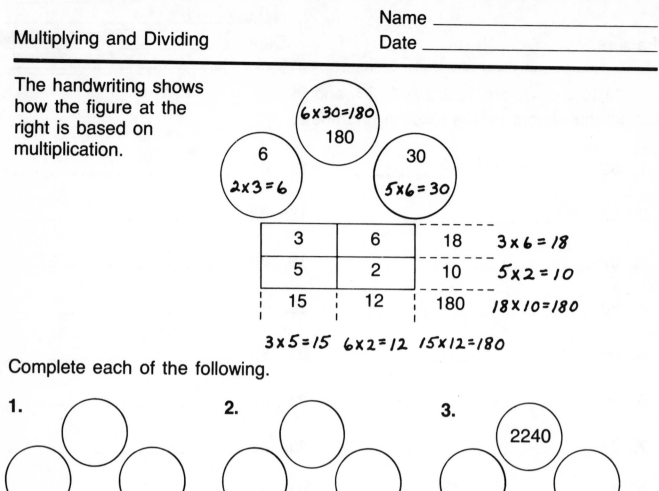

3	6	18	$3 \times 6 = 18$
5	2	10	$5 \times 2 = 10$
15	12	180	$18 \times 10 = 180$

$6 \times 30 = 180$
180
6
$2 \times 3 = 6$
30
$5 \times 6 = 30$

$3 \times 5 = 15$ $6 \times 2 = 12$ $15 \times 12 = 180$

Complete each of the following.

1.

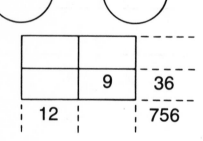

	9	36
12		756

2.

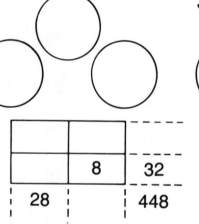

	8	32
28		448

3.

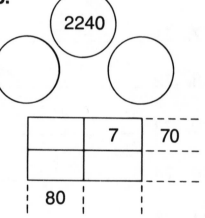

2240

	7	70
80		

4.

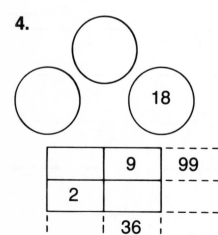

18

	9	99
2		
	36	

5.

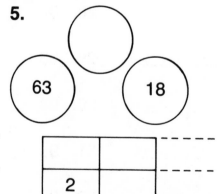

63   18

2		
6		1134

6.

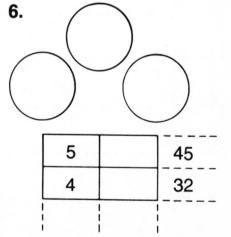

5		45
4		32

Name _____

Date _____

Greatest Common Factor

Find the greatest number that is a factor of both numbers.

1. 15 and 27 __*3*__

2. 24 and 40 _____

3. 36 and 48 _____

4. 49 and 42 _____

5. 34 and 51 _____

6. 45 and 75 _____

7. 62 and 72 _____

8. 26 and 52 _____

9. 62 and 31 _____

10. 43 and 75 _____

11. 36 and 72 _____

12. 52 and 58 _____

13. 63 and 36 _____

14. 81 and 45 _____

15. 40 and 85 _____

16. 26 and 32 _____

17. 77 and 88 _____

18. 28 and 42 _____

19. 53 and 17 _____

20. 66 and 44 _____

The handwriting shows how the figure at the right is based on multiplication.

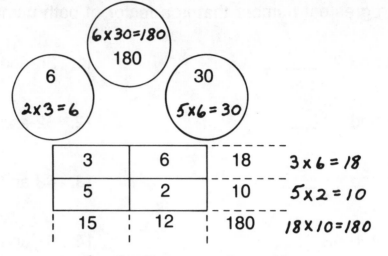

6×30=180
180

6
2×3=6

30
5×6=30

3	6	18	3×6=18
5	2	10	5×2=10
15	12	180	18×10=180

3×5=15   6×2=12   15×12=180

Complete each of the following.

1.

15

	3	
25		675

2.

	9	36
11		
	54	

3.

24

12		72
108		

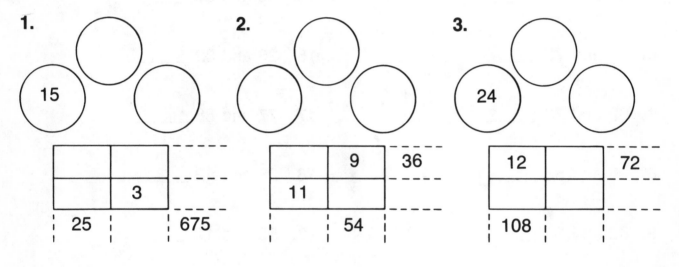

4.

40

5	6	
35		

5.

6		
		40
	32	960

6.

20

5		30
9		
	24	

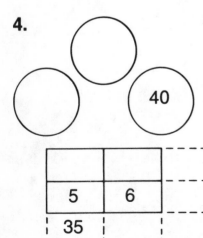
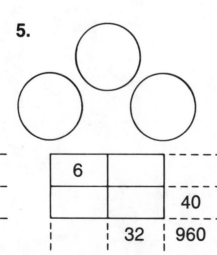
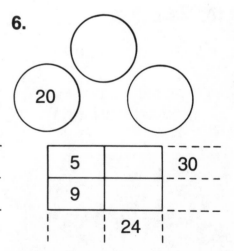

Greatest Common Factor

Name _____
Date _____

Find the greatest number that is a factor of both numbers.

1. 45 and 63 __9__

2. 20 and 50 _____

3. 28 and 35 _____

4. 32 and 64 _____

5. 17 and 34 _____

6. 56 and 84 _____

7. 45 and 73 _____

8. 58 and 87 _____

9. 48 and 80 _____

10. 81 and 99 _____

11. 83 and 14 _____

12. 58 and 29 _____

13. 16 and 64 _____

14. 92 and 46 _____

15. 17 and 34 _____

16. 72 and 54 _____

17. 50 and 75 _____

18. 57 and 42 _____

19. 62 and 93 _____

20. 79 and 82 _____

21. List the numbers between 1 and 100 which have no factors except themselves and 1.

_____

_____

_____

Name _____

Date _____

The handwriting shows how the figure at the right is based on multiplication.

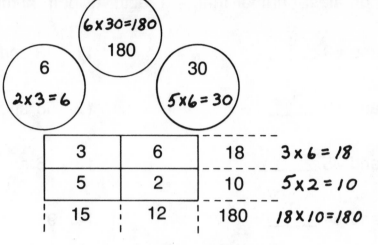

6×30=180
180

6
2×3=6

30
5×6=30

3	6	18	3×6=18
5	2	10	5×2=10
15	12	180	18×10=180

3×5=15 6×2=12 15×12=180

Complete each of the following.

1.

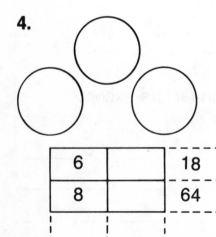

63

		49
6		54

2.

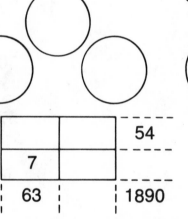

		54
7		
63		1890

3.

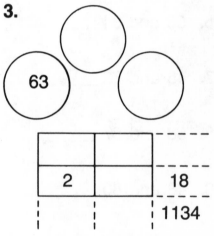

63

2		18
		1134

4.

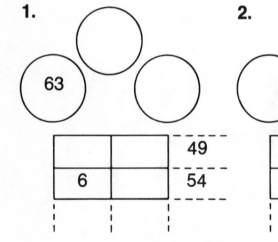

6		18
8		64

5.

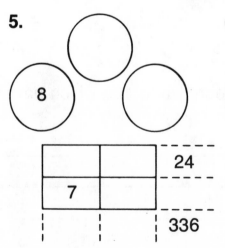

8

		24
7		
		336

6.

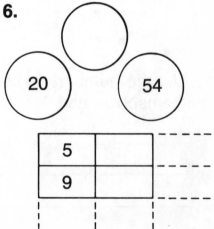

20   54

5		
9		

60

Name _____

Date _____

The prime numbers that are less than 100 are: 2, 3, 5, 7, 11, 13, 17, 19, 23, 29, 31, 37, 41, 43, 47, 53, 59, 61, 67, 71, 73, 79, 83, 89, and 97.

Write each number as a product of prime numbers.

1. $6 =$ _2 x 3_	11. $34 =$ _____
2. $10 =$ _____	12. $57 =$ _____
3. $14 =$ _____	13. $69 =$ _____
4. $15 =$ _____	14. $65 =$ _____
5. $21 =$ _____	15. $30 =$ _____
6. $35 =$ _____	16. $66 =$ _____
7. $62 =$ _____	17. $70 =$ _____
8. $46 =$ _____	18. $42 =$ _____
9. $39 =$ _____	19. $78 =$ _____
10. $55 =$ _____	20. $165 =$ _____

Name _____

Date _____

3^2 means 3×3. The 3 is the *base,* the 2 is the *exponent.* The *exponent* tells how many times the *base* is used as a factor. $3^2 = 3 \times 3 = 9$.

		Base	Exponent	Is read as:	Value
1.	3^2	*3*	*2*	*3 to the second power*	*9*
2.	5^3				
3.	2^5				
4.	4^3				
5.	5^4				
6.	10^5				
7.	11^2				
8.	21^3				

Factoring

The prime numbers that are less than 100 are: 2, 3, 5, 7, 11, 13, 17, 19, 23, 29, 31, 37, 41, 43, 47, 53, 59, 61, 67, 71, 73, 79, 83, 89, and 97.

Write each number as a product of prime numbers.

1. 39 = _3 x 13_	**11.** 183 = _____
2. 95 = _____	**12.** 111 = _____
3. 115 = _____	**13.** 205 = _____
4. 123 = _____	**14.** 213 = _____
5. 161 = _____	**15.** 145 = _____
6. 74 = _____	**16.** 385 = _____
7. 82 = _____	**17.** 561 = _____
8. 159 = _____	**18.** 238 = _____
9. 143 = _____	**19.** 170 = _____
10. 154 = _____	**20.** 345 = _____

Name _____

Date _____

Find the value of each of the following.

1. 2^2 __4__	11. 7^2 _____
2. 2^3 _____	12. 7^3 _____
3. 2^4 _____	13. 8^2 _____
4. 2^5 _____	14. 8^3 _____
5. 2^6 _____	15. 9^2 _____
6. 3^2 _____	16. 10^2 _____
7. 3^3 _____	17. 10^3 _____
8. 3^4 _____	18. 10^4 _____
9. 4^2 _____	19. 10^5 _____
10. 4^3 _____	20. 11^2 _____

The prime numbers that are less than 100 are: 2, 3, 5, 7, 11, 13, 17, 19, 23, 29, 31, 37, 41, 43, 47, 53, 59, 61, 67, 71, 73, 79, 83, 89, and 97.

Write each number as a product of prime numbers.

1. 51 = _3 × 17_	**11.** 301 = _____
2. 133 = _____	**12.** 255 = _____
3. 165 = _____	**13.** 258 = _____
4. 215 = _____	**14.** 273 = _____
5. 159 = _____	**15.** 170 = _____
6. 119 = _____	**16.** 429 = _____
7. 161 = _____	**17.** 114 = _____
8. 138 = _____	**18.** 230 = _____
9. 217 = _____	**19.** 595 = _____
10. 185 = _____	**20.** 451 = _____

Name _____

Date _____

Find the value of each of the following.

1. 2^3 ___8___	**11.** 10^2 _____
2. 3^4 _____	**12.** 10^3 _____
3. 5^2 _____	**13.** 10^4 _____
4. 5^3 _____	**14.** 2×3^2 _____
5. 6^2 _____	**15.** 3×2^3 _____
6. 6^3 _____	**16.** 5×3^2 _____
7. 7^3 _____	**17.** 7×5^2 _____
8. 8^2 _____	**18.** $5^3 \times 2$ _____
9. 9^2 _____	**19.** $3^2 \times 2^3$ _____
10. 9^3 _____	**20.** $5^2 \times 2^2$ _____

Factoring

The prime numbers that are less than 100 are: 2, 3, 5, 7, 11, 13, 17, 19, 23, 29, 31, 37, 41, 43, 47, 53, 59, 61, 67, 71, 73, 79, 83, 89, and 97.

Write each number as a product of prime numbers.

1. 114 = $2 \times 3 \times 19$	**11.** 385 = _____
2. 455 = _____	**12.** 330 = _____
3. 273 = _____	**13.** 856 = _____
4. 230 = _____	**14.** 663 = _____
5. 174 = _____	**15.** 357 = _____
6. 494 = _____	**16.** 1518 = _____
7. 735 = _____	**17.** 715 = _____
8. 138 = _____	**18.** 1045 = _____
9. 165 = _____	**19.** 1309 = _____
10. 231 = _____	**20.** 1085 = _____

Name _____

Date _____

Exponents

Find the value of each of the following.

1. 3^3 _27_	**11.** $2 \times 3 \times 5^2$ _____
2. 4^2 _____	**12.** $3^2 \times 7 \times 11$ _____
3. 5^3 _____	**13.** $2^5 \times 5^2$ _____
4. 7^2 _____	**14.** $5^2 \times 13^2$ _____
5. 8^2 _____	**15.** $7^2 \times 2^5$ _____
6. 9^3 _____	**16.** $2^2 \times 3^3 \times 7$ _____
7. 10^4 _____	**17.** $2^2 \times 7^2$ _____
8. 11^2 _____	**18.** $2^2 \times 3^2 \times 5$ _____
9. 12^3 _____	**19.** $7^2 \times 3^3$ _____
10. 13^2 _____	**20.** $11^2 \times 13^2$ _____

Name _____

Date _____

The prime numbers that are less than 100 are: 2, 3, 5, 7, 11, 13, 17, 19, 23, 29, 31, 37, 41, 43, 47, 53, 59, 61, 67, 71, 73, 79, 83, 89, and 97.

Write each number as a product of prime numbers.

1. 51 = __3 X 17__	**11.** 183 = _____
2. 95 = _____	**12.** 258 = _____
3. 115 = _____	**13.** 208 = _____
4. 230 = _____	**14.** 213 = _____
5. 159 = _____	**15.** 145 = _____
6. 74 = _____	**16.** 429 = _____
7. 161 = _____	**17.** 561 = _____
8. 158 = _____	**18.** 231 = _____
9. 165 = _____	**19.** 170 = _____
10. 153 = _____	**20.** 345 = _____

Exponents

Find the value of each of the following.

1. 2^3 ___8___	**11.** $2 \times 3^2 \times 5$ _____
2. 2^5 _____	**12.** $3 \times 5 \times 7^2$ _____
3. 3^2 _____	**13.** $2^3 \times 3 \times 5$ _____
4. 5^2 _____	**14.** $5^2 \times 3^3$ _____
5. 5^3 _____	**15.** $2^2 \times 13^2$ _____
6. 7^2 _____	**16.** 7×11^2 _____
7. 11^2 _____	**17.** $5^2 \times 17$ _____
8. 13^2 _____	**18.** $2^4 \times 29$ _____
9. 17^2 _____	**19.** $3^2 \times 5^3$ _____
10. 19^2 _____	**20.** $3^4 \times 7$ _____

Products of Whole Numbers

Name _____

Date _____

Find the products.

1. $\begin{array}{r} 3,040 \\ \times\ \ \ 7 \\ \hline 21,280 \end{array}$	**2.** $\begin{array}{r} 4,070 \\ \times\ \ \ 6 \\ \hline \end{array}$	**3.** $\begin{array}{r} 6,080 \\ \times\ \ \ 5 \\ \hline \end{array}$
4. $\begin{array}{r} 50,700 \\ \times\ \ \ 8 \\ \hline \end{array}$	**5.** $\begin{array}{r} 40,030 \\ \times\ \ \ 9 \\ \hline \end{array}$	**6.** $\begin{array}{r} 27 \\ \times\ 3 \\ \hline \end{array}$
7. $\begin{array}{r} 763 \\ \times\ \ 6 \\ \hline \end{array}$	**8.** $\begin{array}{r} 304 \\ \times\ \ 7 \\ \hline \end{array}$	**9.** $\begin{array}{r} 6,500 \\ \times\ \ \ 5 \\ \hline \end{array}$
10. $\begin{array}{r} 5,957 \\ \times\ \ \ 8 \\ \hline \end{array}$	**11.** $\begin{array}{r} 6,000 \\ \times\ \ \ 3 \\ \hline \end{array}$	**12.** $\begin{array}{r} 3,009 \\ \times\ \ \ 9 \\ \hline \end{array}$

Products of Whole Numbers

Find the products.

1.   2,718
  ×   5
  13,590
	2. 5,380
	  × 7
		3. 16,000
		  × 6

4.   5,923
  ×   6
	5. 8,323
	  × 3
		6. 41,123
		  × 8

7.   24
  ×22
	8. 35
	  ×31
		9. 43
		  ×25

10.   29
  ×31
	11. 54
	  ×42
		12. 62
		  ×17

Products of Whole Numbers

Find the products.

1. $\begin{array}{r} 3{,}224 \\ \times\quad 8 \\ \hline 25{,}792 \end{array}$	**2.** $\begin{array}{r} 5{,}112 \\ \times\quad 7 \\ \hline \end{array}$	**3.** $\begin{array}{r} 6{,}051 \\ \times\quad 9 \\ \hline \end{array}$
4. $\begin{array}{r} 5{,}621 \\ \times\quad 14 \\ \hline \end{array}$	**5.** $\begin{array}{r} 8{,}035 \\ \times\quad 23 \\ \hline \end{array}$	**6.** $\begin{array}{r} 9{,}221 \\ \times\quad 32 \\ \hline \end{array}$
7. $\begin{array}{r} 7{,}005 \\ \times\quad 15 \\ \hline \end{array}$	**8.** $\begin{array}{r} 4{,}331 \\ \times\quad 19 \\ \hline \end{array}$	**9.** $\begin{array}{r} 5{,}662 \\ \times\quad 26 \\ \hline \end{array}$
10. $\begin{array}{r} 5{,}162 \\ \times\quad 37 \\ \hline \end{array}$	**11.** $\begin{array}{r} 8{,}092 \\ \times\quad 25 \\ \hline \end{array}$	**12.** $\begin{array}{r} 3{,}667 \\ \times\quad 62 \\ \hline \end{array}$

Products of Whole Numbers

Name _____

Date _____

Find the products.

1. $\quad$ 3,006 $\times \quad$ 61 $\overline{\quad 3006}$ $\underline{18036}$ $\overline{183,366}$	**2.** $\quad$ 4,105 $\times \quad$ 52	**3.** $\quad$ 9,113 $\times \quad$ 81
4. $\quad$ 6,112 $\times \quad$ 54	**5.** $\quad$ 8,235 $\times \quad$ 43	**6.** $\quad$ 4,332 $\times \quad$ 91
7. $\quad$ 4,366 $\times \quad$ 56	**8.** $\quad$ 4,717 $\times \quad$ 43	**9.** $\quad$ 5,324 $\times \quad$ 18
10. $\quad$ 9,111 $\times \quad$ 43	**11.** $\quad$ 5,443 $\times \quad$ 22	**12.** $\quad$ 8,556 $\times \quad$ 62

Products of Whole Numbers

Name _____

Date _____

Find the products.

1. 5,117 × 23 15351 10234 117,691	**2.** 3,452 × 41	**3.** 4,002 × 35
4. 6,105 × 43	**5.** 6,331 × 36	**6.** 3,125 × 24
7. 7,343 × 55	**8.** 9,222 × 72	**9.** 5,317 × 63
10. 8,105 × 30	**11.** 6,347 × 50	**12.** 7,334 × 90

75

Products of Whole Numbers

Name _____

Date _____

Answer each question.

1. What will be the cost 35 math books if they
   cost $5 each? $175

   $$35 \times \$5 = \$175$$

2. If there are 5280 feet in a mile, how many feet
   are there in 17 miles?

3. How far will Jim travel in 9 hours if he is
   driving an automobile at an average rate of
   55 miles per hour?

4. If there is an average of eleven words on each
   line of a page in a certain book, how many
   words are on a page containing 44 lines?

5. How many days are there in fifteen years if
   there are 365 days in each year?

6. There are 36 inches in a yard. How many
   inches are there in 25 yards?

Products of Whole Numbers

Name _____

Date _____

Find the products.

1. 6,335 × 16 38010 6335 101,360	**2.** 3,173 × 22	**3.** 2,631 × 43
4. 5,535 × 47	**5.** 8,023 × 62	**6.** 9,115 × 93
7. 8,231 × 46	**8.** 4,034 × 35	**9.** 7,662 × 54
10. 693 ×405	**11.** 752 ×802	**12.** 834 ×706

77

Name _____

Date _____

Answer each question.

1. If blue jeans are on sale for $13 a pair, how much will 25 pairs sell for? $325

   $13 x 25 = $325

2. There are 100 centimeters in a meter. How many centimeters are there in 726 meters?

3. If an airplane is traveling at an average rate of 625 miles an hour, how far will the plane travel in 8 hours?

4. If there is an average of 26 books on each of 15 bookshelves, how many books are on the shelves?

5. How many weeks are there in 17 years if there are 52 weeks in a year?

6. There are 1000 meters in a kilometer. How many meters are there in 35 kilometers?

Products of Whole Numbers

Find the products.

1. 5,112 × 13 _15 3 3 6_ _5 1 1 2_ _66,456_	**2.** 4,052 × 25	**3.** 6,334 × 43
4. 6,327 × 16	**5.** 8,135 × 32	**6.** 4,102 × 17
7. 9,362 × 38	**8.** 4,913 × 55	**9.** 8,716 × 40
10. 4,323 × 63	**11.** 7,654 × 57	**12.** 5,032 × 49

79

Name _____

Date _____

Answer each question.

1. If automobile tires cost $43 each, what would be the cost of 5 tires? $215

$43 × 5 = $215

2. There are 36 inches in a yard. How many inches are there in 50 yards?

3. How many miles will an automobile travel in 12 hours if it is traveling at an average speed of 52 miles per hour?

4. There are twelve pencils in a box and there are 144 boxes in a gross. How many pencils are there in a gross?

5. Using 365 days in a year, how many days has a person lived when 12 years old?

6. There are 10 decimeters in a meter. How many decimeters are there in 352 meters?

Factoring of Whole Numbers

Name _____

Date _____

List the prime numbers that are less than 50.

Write each number as a product of prime numbers. Use exponents whenever possible.

1. 28 $2^2 \times 7$	**11.** 96 _____
2. 85 _____	**12.** 169 _____
3. 60 _____	**13.** 268 _____
4. 112 _____	**14.** 184 _____
5. 115 _____	**15.** 142 _____
6. 183 _____	**16.** 279 _____
7. 190 _____	**17.** 260 _____
8. 1225 _____	**18.** 182 _____
9. 280 _____	**19.** 456 _____
10. 255 _____	**20.** 188 _____

Place Value

Write the number.

1. one hundred eighty-five ___*185*___

2. six hundred sixty-two _____

3. eight hundred twenty-nine _____

4. nine hundred ninety-eight _____

5. four hundred seventy-four _____

6. five hundred eight _____

7. two thousand six hundred twenty _____

8. five thousand two hundred nine _____

9. seven thousand two hundred two _____

10. six thousand two hundred _____

Build a number. Write each digit in place. Put in zeros if you need them.

	thousands	hundreds	tens	ones
11. 5 in the tens place, 8 in the ones place	____	____	____	____
12. 6 in the ones place, 4 in the hundreds place	____	____	____	____
13. 1 in the hundreds place, 5 in the ones place, 6 in the tens place	____	____	____	____
14. 8 in the thousands place, 1 in the tens place, 9 in the hundreds place	____	____	____	____

Factoring of Whole Numbers

Name _____

Date _____

List the prime numbers that are less than 50.

Write each number as a product of prime numbers. Use exponents whenever possible.

1. 20 $2^2 \times 5$	**11.** 80 _____
2. 63 _____	**12.** 189 _____
3. 40 _____	**13.** 248 _____
4. 105 _____	**14.** 136 _____
5. 104 _____	**15.** 140 _____
6. 153 _____	**16.** 255 _____
7. 175 _____	**17.** 250 _____
8. 1001 _____	**18.** 190 _____
9. 306 _____	**19.** 496 _____
10. 352 _____	**20.** 187 _____

Place Value

Write the number.

1. one hundred fifty-seven ___*157*___

2. four hundred thirty-two _____

3. five hundred seventy-five _____

4. eight hundred eighty-one _____

5. seven hundred twenty-six _____

6. four hundred seven _____

7. one thousand three hundred thirty _____

8. four thousand one hundred seventeen _____

9. three thousand six hundred fifteen _____

10. four thousand eight hundred _____

Build a number. Write each digit in place. Put in zeros if you need them.

	thousands	hundreds	tens	ones
11. 4 in the tens place, 7 in the ones place	_____	_____	_____	_____
12. 9 in the ones place, 3 in the tens place	_____	_____	_____	_____
13. 2 in the hundreds place, 6 in the ones place, 4 in the tens place	_____	_____	_____	_____
14. 6 in the thousands place, 1 in the tens place, 2 in the ones place, 1 in the hundreds place	_____	_____	_____	_____

Factoring of Whole Numbers

Name _____

Date _____

List the prime numbers that are less than 50.

Write each number as a product of prime numbers. Use exponents whenever possible.

1. 575 5^2x 23	**11.** 188 _____
2. 63 _____	**12.** 189 _____
3. 455 _____	**13.** 1450 _____
4. 1044 _____	**14.** 637 _____
5. 344 _____	**15.** 244 _____
6. 999 _____	**16.** 171 _____
7. 1040 _____	**17.** 370 _____
8. 2009 _____	**18.** 204 _____
9. 475 _____	**19.** 175 _____
10. 248 _____	**20.** 208 _____

Place Value

Write the number.

1. one hundred eighty-two _____*182*_____

2. three hundred seventy-six _____

3. four hundred three _____

4. six hundred twenty-five _____

5. four hundred thirty-four _____

6. three hundred twenty-one _____

7. one thousand six hundred ninety _____

8. four thousand ninety-six _____

9. six thousand one hundred seventeen _____

10. two thousand nine hundred thirty _____

Build a number. Write each digit in place. Put in zeros if you need them.

	thousands	hundreds	tens	ones
11. 7 in the tens place, 5 in the ones place	_____	_____	_____	_____
12. 8 in the ones place, 6 in the tens place	_____	_____	_____	_____
13. 2 in the ones place, 1 in the hundreds place	_____	_____	_____	_____
14. 1 in the thousands place, 4 in the tens place, 3 in the ones place	_____	_____	_____	_____

Factoring of Whole Numbers

List the prime numbers that are less than 50.

Write each number as a product of prime numbers. Use exponents whenever possible.

1. 132 $2^2 \times 3 \times 11$	**11.** 246 _____
2. 615 _____	**12.** 435 _____
3. 222 _____	**13.** 440 _____
4. 296 _____	**14.** 480 _____
5. 117 _____	**15.** 128 _____
6. 812 _____	**16.** 333 _____
7. 282 _____	**17.** 114 _____
8. 414 _____	**18.** 480 _____
9. 850 _____	**19.** 140 _____
10. 117 _____	**20.** 243 _____

Place Value

Name _____
Date _____

Write the number.

1. one hundred ninety-four ___194___

2. two hundred ninety-seven _____

3. two hundred sixty-seven _____

4. eight hundred nine _____

5. seven hundred twenty-six _____

6. six hundred three _____

7. one thousand seven hundred thirty _____

8. three thousand six hundred twelve _____

9. three thousand ninety-two _____

10. six thousand one hundred ten _____

Build a number. Write each digit in place. Put in zeros if you need them.

	thousands	hundreds	tens	ones
11. 9 in the tens place, 9 in the ones place	_____	_____	_____	_____
12. 7 in the ones place, 4 in the hundreds place	_____	_____	_____	_____
13. 4 in the thousands place, 3 in the hundreds place, 2 in the ones place	_____	_____	_____	_____
14. 9 in the thousands place, 7 in the tens place, 6 in the ones place	_____	_____	_____	_____

Factoring of Whole Numbers

Name _____

Date _____

List the prime numbers that are less than 50.

Write each number as a product of prime numbers. Use exponents whenever possible.

1. 56 $\underline{2^3 \times 7}$	**11.** 150 _____
2. 325 _____	**12.** 217 _____
3. 90 _____	**13.** 165 _____
4. 88 _____	**14.** 182 _____
5. 1127 _____	**15.** 385 _____
6. 243 _____	**16.** 208 _____
7. 282 _____	**17.** 441 _____
8. 250 _____	**18.** 136 _____
9. 1573 _____	**19.** 775 _____
10. 90 _____	**20.** 820 _____

Place Value

Name _____
Date _____

Write the number.

1. one hundred twenty-six __126__

2. four hundred thirty-two _____

3. six hundred ninety-two _____

4. four hundred thirty-three _____

5. seven hundred two _____

6. eight hundred twenty-one _____

7. one thousand eight hundred eighty _____

8. four thousand one hundred seventeen _____

9. eight thousand two hundred twenty _____

10. four thousand forty-one _____

Build a number. Write each digit in place. Put in zeros if you need them.

	thousands	hundreds	tens	ones
11. 2 in the ones place, 8 in the tens place	____	____	____	____
12. 4 in the tens place, 7 in the ones place	____	____	____	____
13. 1 in the tens place, 3 in the thousands place, 4 in the ones place	____	____	____	____
14. 8 in the thousands place, 1 in the ones place, 1 in the tens place	____	____	____	____

Division of Whole Numbers

Name _____

Date _____

Divide.

1. 2394 ÷ 7 $$\begin{array}{r} 342 \\ 7\overline{)2394} \\ 21 \\ \hline 29 \\ 28 \\ \hline 14 \\ 14 \\ \hline 0 \end{array}$$	**2.** 4465 ÷ 5	**3.** 1863 ÷ 3
4. 6976 ÷ 8	**5.** 59,229 ÷ 9	**6.** 48,447 ÷ 7
7. 199,120 ÷ 5	**8.** 30,745 ÷ 43	**9.** 23,733 ÷ 81

Name _____

Date _____

Solve each problem.

1. If Jim drives his automobile for 8 hours and travels 416 miles, what is his average rate of speed? *52 miles per hour*

   416 ÷ 8 = 52

2. If an automobile mechanic makes $15,360 a year, what is his monthly salary?

3. If one can of paint covers 15 square feet, how many cans of paint will it take to paint 945 square feet?

4. An assembly line at a bottling plant fills 13 cases of bottles per minute. How long will it take to fill 754 cases?

5. If the total cost of a picnic for 37 people is $111, what is the cost for each person?

Division of Whole Numbers

Name _____

Date _____

Divide.

1. $1928 \div 2$	2. $3284 \div 4$	3. $6835 \div 5$

$$\begin{array}{r} 964 \\ 2\overline{)1928} \\ 18 \\ \hline 12 \\ 12 \\ \hline 8 \\ 8 \\ \hline 0 \end{array}$$

4. $40{,}296 \div 8$	5. $30{,}462 \div 6$	6. $804{,}168 \div 9$

7. $32{,}025 \div 75$	8. $4032 \div 32$	9. $30{,}681 \div 63$

Name _____

Date _____

Solve each problem.

---

1. An airplane flies 2850 miles in 6 hours. What is the average speed in miles per hour? *475 miles per hour*

$$2850 \div 6 = 475$$

---

2. If a telephone installer earns a salary of $15,600 a year, what is her monthly salary?

---

3. If 1 square yard of carpeting costs $13, how many square yards of floor can be carpeted for $585?

---

4. A cannery produces 22 cases of peaches every hour. How many hours will it take to produce 528 cases?

---

5. If the total cost of a concert for 525 people is $3150, what is the cost for each person?

Name _____

Date _____

Divide.

1. 9580 ÷ 4	**2.** 27,764 ÷ 4	**3.** 490,232 ÷ 8
4. 285,544 ÷ 7	**5.** 357,744 ÷ 6	**6.** 240,006 ÷ 6
7. 1404 ÷ 52	**8.** 693 ÷ 21	**9.** 2795 ÷ 43

For problem 1, the worked long division:

$$\begin{array}{r} 2395 \\ 4\overline{)9580} \\ 8 \\ \hline 15 \\ 12 \\ \hline 38 \\ 36 \\ \hline 20 \\ 20 \\ \hline 0 \end{array}$$

Name _____

Date _____

Solve each problem.

1. The Bank of America building in San
   Francisco is approximately 780 feet tall and
   is 52 stories high. What is the average height
   of each floor?   *15 feet*

   $780 \div 52 = 15$

2. A speed skater can skate 3000 meters in 6
   minutes. How many meters is this per
   minute?

3. A truckload of grain with a total of 30,561
   pounds contains 501 bushels. What is the
   weight of each bushel?

4. After traveling 315 miles it takes 15 gallons
   of gasoline to refill the tank. What was the
   average gasoline mileage?

5. If a bricklayer makes $14,784 in a year, what
   is his average monthly wage?

Division of Whole Numbers

Divide.

1. 249,735 ÷ 3	

$$\begin{array}{r} 83,245 \\ 3\overline{)249,735} \\ \underline{24} \\ 9 \\ \underline{9} \\ 7 \\ \underline{6} \\ 13 \\ \underline{12} \\ 15 \\ \underline{15} \\ 0 \end{array}$$

2. 400,085 ÷ 5

3. 582,099 ÷ 7

4. 3901 ÷ 47

5. 1656 ÷ 24

6. 5986 ÷ 82

7. 50,406 ÷ 62

8. 13,804 ÷ 29

9. 12,054 ÷ 42

97

Division Problems

Solve each problem.

1. A load of onions containing 127 bags weighs
   a total of 5334 pounds. How much does each
   bag of onions weigh?   *42 pounds*

   $5334 \div 127 = 42$

2. If a sheet-metal worker earns $13,884 in a
   year, what is his average monthly salary?

3. The Transamerica Building in San Francisco
   is approximately 864 feet high and is 48
   stories tall. What is the average height of
   each floor?

4. If there are 16,000 acres of land in 25 sections
   of land, how many acres are there in one
   section?

5. An automobile trip of 1012 miles took 22
   hours of driving time. What was the average
   rate of travel?

98

Division of Whole Numbers

Name _____

Date _____

Divide.

1. $21{,}560 \div 5$	2. $19{,}308 \div 6$	3. $30{,}568 \div 8$

$$
\begin{array}{r}
4{,}312 \\
5\overline{)21{,}560} \\
20 \\
\hline
15 \\
15 \\
\hline
6 \\
5 \\
\hline
10 \\
10 \\
\hline
0
\end{array}
$$

4. $42{,}228 \div 9$	5. $1334 \div 29$	6. $1725 \div 23$

7. $3612 \div 86$	8. $18{,}228 \div 28$	9. $27{,}531 \div 63$

Division Problems

Name _____

Date _____

Solve each problem.

1. The average milk consumption for each person in the United States during the year of 1974 was approximately 240 pounds. If milk weighs 8 pounds per gallon, how many gallons did each person consume? *30 gallons*

   $240 \div 8 = 30$

2. An airplane flight of 5016 miles took total flying time of 12 hours. What was the average rate of travel?

3. If California has a population of approximately 20,960,000 people and a land area of approximately 160,000 square miles, what is the average population per square mile?

4. If a nurse earns $15,360 per year, what is her monthly salary?

5. A family of 4 people uses 15,000 gallons of water in a month. If there were 30 days in a month, how many gallons is this per day?

   On the average, how much water does each person use each day?

ANSWERS

Page 1 **1.** 26 **2.** 197 **3.** 119 **4.** 1369 **5.** 758 **6.** 928 **7.** 2,265,648 **8.** 770,861 **9.** 394,644 **10.** 1,189,767 **11.** 68 **12.** 115 **13.** 589 **14.** 3328 **15.** 1529 **16.** 13,723 **17.** 63 **18.** 1172 **19.** 89,376 **20.** 265 **21.** 2321 **22.** 116,213

Page 2 **1.** 262 **2.** 684 **3.** 399 **4.** 675 **5.** 178 **6.** 208` **7.** 33,431 **8.** 51,125 **9.** 58,276 **10.** 67,949 **11.** 304,773 **12.** 358,239 **13.** 359,886 **14.** 606,655 **15.** 597 **16.** 2177 **17.** 608 **18.** 111

Page 3 **1.** 30 **2.** 306 **3.** 87 **4.** 2050 **5.** 2412 **6.** 826 **7.** 7,165,148 **8.** 1,374,855 **9.** 8,370,381 **10.** 3,087,882 **11.** 225 **12.** 130 **13.** 1252 **14.** 1038 **15.** 4268 **16.** 14,935 **17.** 182 **18.** 1708 **19.** 67,404 **20.** 982 **21.** 1145 **22.** 137,017

Page 4 **1.** 815 **2.** 348 **3.** 298 **4.** 364 **5.** 278 **6.** 407 **7.** 64,739 **8.** 93,586 **9.** 46,391 **10.** 79,669 **11.** 800,703 **12.** 872,701 **13.** 123,322 **14.** 849,267 **15.** 112 **16.** 1298 **17.** 189 **18.** 665

Page 5 **1.** 29 **2.** 345 **3.** 136 **4.** 2287 **5.** 2052 **6.** 1893 **7.** 7,258,242 **8.** 991,210 **9.** 874,365 **10.** 10,885,258 **11.** 101 **12.** 188 **13.** 426 **14.** 2688 **15.** 1202 **16.** 9823 **17.** 442 **18.** 891 **19.** 114,957 **20.** 1488 **21.** 160 **22.** 58,394

Page 6 **1.** 157 **2.** 617 **3.** 289 **4.** 811 **5.** 181 **6.** 53 **7.** 96,593 **8.** 92,288 **9.** 47,432 **10.** 35,929 **11.** 477,442 **12.** 601,971 **13.** 164,161 **14.** 319,019 **15.** 741 **16.** 3344 **17.** 902 **18.** 763

Page 7 **1.** 23 **2.** 253 **3.** 222 **4.** 2160 **5.** 1529 **6.** 1862 **7.** 2,962,141 **8.** 432,053 **9.** 657,137 **10.** 13,893,538 **11.** 188 **12.** 195 **13.** 1105 **14.** 1637 **15.** 1209 **16.** 12,784 **17.** 1044 **18.** 1118 **19.** 111,409 **20.** 236 **21.** 1625 **22.** 103,143

Page 8 **1.** 754 **2.** 658 **3.** 9 **4.** 862 **5.** 57 **6.** 518 **7.** 57,748 **8.** 65,085 **9.** 52,190 **10.** 44,279 **11.** 101,677 **12.** 810,009 **13.** 1308 **14.** 789,111 **15.** 267 **16.** 4497 **17.** 1520 **18.** 775

Page 9 **1.** 28 **2.** 315 **3.** 186 **4.** 2083 **5.** 1708 **6.** 1587 **7.** 6,674,469 **8.** 1,134,706 **9.** 438,867 **10.** 12,307,683 **11.** 218 **12.** 199 **13.** 972 **14.** 6393 **15.** 1746 **16.** 8427 **17.** 78 **18.** 1627 **19.** 137,525 **20.** 616 **21.** 1429 **22.** 93,080

Page 10 **1.** 829 **2.** 330 **3.** 87 **4.** 859 **5.** 226 **6.** 85 **7.** 48,879 **8.** 34,869 **9.** 55,738 **10.** 37,770 **11.** 375,588 **12.** 892,389 **13.** 224,277 **14.** 237,779 **15.** 844 **16.** 1378 **17.** 455 **18.** 581

Page 11 **1.** 6, 2 **2.** 18, 0 **3.** 12, 4 **4.** 8, 2 **5.** 12, 2 **6.** 14, 2 **7.** 3, 1 **8.** 8, 8 **9.** 10, 8 **10.** 11, 1 **11.** 13, 1 **12.** 11, 5 **13.** 0, 0 **14.** 17, 1 **15.** 9, 3 **16.** 9, 1 **17.** 16, 2 **18.** 8, 0 **19.** 16, 0 **20.** 14, 4

Page 12 Starting with the row of circles, missing numbers are given row by row from left to right. **1.** 79, 189, 110; 63; 126; 105, 84, 189 **2.** 1385, 2900, 1515; 1538; 1362; 195, 2705, 2900 **3.** 1298, 2405, 1107; 2259; 146; 1360, 1045, 2405 **4.** 161, 262, 101; 65; 36; 125, 137, 262 **5.** 452, 272; 93; 157, 87, 244; 202, 452 **6.** 206, 463; 155; 102, 5, 107; 303, 160

Page 13 **1.** 20, 6 **2.** 18, 14 **3.** 27, 3 **4.** 19, 3 **5.** 22, 18 **6.** 15, 13 **7.** 29, 7 **8.** 14, 14 **9.** 37, 1 **10.** 29, 5 **11.** 24, 4 **12.** 22, 20 **13.** 25, 7 **14.** 21, 13 **15.** 18, 4 **16.** 21, 11 **17.** 29, 11 **18.** 21, 5 **19.** 20, 16 **20.** 16, 4

Page 14 Starting with the row of circles, missing numbers are given row by row from left to right. **1.** 820, 1933, 1113; 975; 958; 1084, 849, 1933 **2.** 878, 5131, 4253; 3545; 1586; 940, 4191, 5131 **3.** 1430, 6116, 4686; 2019; 4097; 4369, 1747, 6116 **4.** 612, 1262, 650; 151; 499, 653; 305, 1262 **5.** 1052, 620; 211, 148, 359; 472; 683, 369 **6.** 427; 290, 180, 470; 137; 515, 317, 832

Page 15 **1.** 72, 28 **2.** 50, 4 **3.** 60, 16 **4.** 66, 14 **5.** 60, 12 **6.** 110, 56 **7.** 73, 27 **8.** 50, 6 **9.** 60, 6 **10.** 63, 17 **11.** 60, 8 **12.** 90,42 **13.** 40, 16 **14.** 70,22 **15.** 40, 4 **16.** 50, 2 **17.** 77, 23 **18.** 36, 16 **19.** 82, 38 **20.** 40, 12

Page 16 Starting with the top row of circles, missing numbers are given row by row from left to right. **1.** 939, 1483, 544; 745; 738; 711, 772, 1483 **2.** 820, 8306, 7486; 6480; 1826; 1264, 7042, 8306 **3.** 1159, 7523, 6364; 6843; 680; 1285, 6238, 7523 **4.** 1035, 681; 142, 343; 480, 212; 622, 413 **5.** 570, 322; 144; 178, 271; 655, 237, 892 **6.** 689, 957, 268; 627; 153, 215; 780, 177

Page 17 **1.** 90, 44 **2.** 48, 8 **3.** 75, 25 **4.** 80, 22 **5.** 100, 44 **6.** 70, 18 **7.** 70, 26 **8.** 88, 32 **9.** 60, 10 **10.** 74, 26 **11.** 56, 4 **12.** 40, 6 **13.** 107, 53 **14.** 80, 36 **15.** 70, 20 **16.** 120, 66 **17.** 44, 4 **18.** 70, 14 **19.** 110, 54 **20.** 87, 33

Page 18 Starting with the top row of circles, missing numbers are given row by row from left to right. **1.** 1225, 2172, 947; 1204; 968; 1189, 983, 2172 **2.** 3939, 5292, 1353; 2181; 3111; 2260, 3032, 5292 **3.** 4307, 5623, 1316; 4186; 1437; 3692, 1931, 5623 **4.** 1580, 2958, 1378; 889, 458; 920, 1611; 1149 **5.** 1170; 399, 386, 785; 385; 650, 520, 1170 **6.** 613, 254; 169; 85, 266; 517, 350, 867

Page 19 **1.** 12, 2 **2.** 8, 0 **3.** 20, 6 **4.** 21, 13 **5.** 29, 11 **6.** 31, 9 **7.** 27, 9 **8.** 33, 9 **9.** 18, 10 **10.** 34, 2 **11.** 27, 11 **12.** 25, 11 **13.** 30, 12 **14.** 80, 30 **15.** 66, 14 **16.** 40, 4 **17.** 70, 18 **18.** 75, 25 **19.** 80, 36 **20.** 110,54

Page 20 Starting with the top row of circles, missing numbers are given row by row from left to right. **1.** 107, 241, 134; 110; 131; 122, 119, 241 **2.** 1460, 3241, 1781; 1891; 1350; 300, 2941, 3241 **3.** 1321, 5901, 4580; 2004; 3897; 4280, 1621, 5901 **4.** 1035, 681; 142, 343; 480, 212; 622, 413 **5.** 427; 290, 180, 470; 137; 515, 317, 832 **6.** 613, 254; 169; 85, 266; 517, 350, 867

Page 21 **1.** the final digit of the number must be 0,2,4,6, or 8 **2.** the sum of the digits of the number is divisible by 3 **3.** the final digit of the number must be 0 or 5 **4.** 102, 104, 106, 108, 110, 112, 114, 116, 118, 120, 122, 124 **5.** 108, 111, 114, 117, 120, 123, 126, 129, 132, 135, 138, 141 **6.** 105, 110, 115, 120, 125, 130, 135, 140, 145, 150, 155, 160

Page 22 **1.** 16, 186, 150, 296, 426, 810 **2.** 72, 63, 87, 114, 363, 825 **3.** 85, 125, 160, 180, 295, 480, 325 365 **4.** 90, 156, 390, 516, 402 **5.** 250, 160, 310, 220, 400, 380 **6.** 630, 300

Page 23 **1.** 152, 154, 156, 158, 160, 162, 164, 166, 168, 170, 172, 174 **2.** 153, 156, 159, 162, 165, 168, 171, 174, 177, 180, 183, 186 **3.** 155, 160, 165, 170, 175, 180, 185, 190, 195, 200, 205, 210 **4.** 105, 112, 119, 126, 133, 140, 147, 154, 161, 168, 175, 182 **5.** 276, 828, 322, 810, 832, 944, 500 **6.** 372, 417, 456, 915, 189, 324, 702

Page 24 **1.** 255, 830, 505, 700, 535, 960 **2.** 28, 35, 49, 70, 63, 14, 91, 98 **3.** 342, 834, 618, 822, 516 **4.** 160, 700, 470, 620, 900 **5.** 255, 150, 210, 375, 345, 105, 810 **6.** 42, 28, 98 **7.** 35, 70, 105

Page 25 **1.** 202, 204, 206, 208, 210, 212, 214, 216, 218, 220, 222, 224 **2.** 201, 204, 207, 210, 213, 216, 219, 222, 225, 228, 231, 234 **3.** 205, 210, 215, 220, 225, 230, 235, 240, 245, 250, 255, 260 **4.** 104, 108, 112, 116, 120, 124, 128, 132, 136, 140, 144, 148 **5.** the last two digits considered as a number are divisible by 4 **6.** 203, 210, 217, 224, 231, 238, 245, 252, 259, 266, 273, 280

Page 26 **1.** 324, 234, 618, 432, 900 **2.** 42, 70, 84, 14, 98 **3.** 42, 84, 126, 63, 105 **4.** 70, 105, 210, 280, 140 **5.** 132, 156, 116, 180, 112 **6.** 6 **7.** 14 **8.** 10 **9.** 15 **10.** 21

Page 27 **1.** 38, 62, 84, 30, 54, 32, 94 **2.** 45, 84, 30, 39, 54, 63 **3.** 25, 45, 30, 55 **4.** 84, 91, 63

Page 28 **1.** 70, 168, 182, 330, 154, 162, 234, 304 **2.** 195, 168, 330, 162, 234, 183 **3.** 70, 195, 330 **4.** 70, 168, 182, 154 **5.** 204, 208, 212, 216, 220, 224, 228, 232, 236, 240, 244, 248 **6.** 216, 228, 236, 292, 284, 296, 264, 272, 220, 300, 312, 532, 716, 436, 200, 840

Page 29 **1.** 38, 62, 84, 126, 30, 76, 142, 96, 130, 156, 70, 182, 330, 154, 162, 180, 250, 210, 200, 196, 304, 434 **2.** 45, 84, 126, 30, 96, 156, 195, 330, 162, 180, 210, 327, 357 **3.** 84, 76, 96, 156, 180, 200, 196, 304 **4.** 25, 45, 30, 130, 70, 195, 330, 235, 180, 250, 210, 200 **5.** 84, 126, 70, 182, 154, 210, 196, 434, 357, 343 **6.** 84, 126, 30, 96, 156, 330, 162, 180, 210

Page 30 **1.** 30, 130, 70, 330, 180, 250, 210, 200 **2.** 84, 126, 70, 182, 154, 210, 196, 434 **3.** 45, 30, 195, 330, 180, 210 **4.** 84, 126, 210, 357 **5.** 70, 210 **6.** 180, 200

Pages 31-40 Starting with the top row of circles, missing numbers are given row by row from left to right.

Page 31 **1.** 25, 1200, 48; 40; 30; 30, 40, 1200 **2.** 8, 448, 56; 28; 16; 32, 14, 448 **3.** 63, 1701, 27; 81; 21; 27, 63, 1701 **4.** 36, 1296, 36; 36; 36; 16, 81, 1296 **5.** 36, 2268, 63; 42; 54; 54, 42, 2268 **6.** 80, 1200, 15; 24; 50; 40, 30, 1200

Page 32 **1.** 42, 336, 8; 12; 28; 24, 14, 336 **2.** 70, 2800, 40; 35; 80; 56, 50, 2800 **3.** 88, 2376, 27; 72; 33; 24, 99, 2376 **4.** 30, 600, 20; 24; 25; 30, 20, 600 **5.** 56, 672, 12; 14; 48; 42, 16, 672 **6.** 63, 756, 12; 27; 28; 36, 21, 756

Page 33 **1.** 35, 840, 24; 42; 20; 28, 30, 840 **2.** 21, 378, 18; 27; 14; 6, 63, 378 **3.** 48, 576, 12; 32; 18; 24, 24, 576 **4.** 66, 528, 8; 44; 12; 22, 24, 528 **5.** 30, 1050, 35; 21; 50; 15, 70, 1050 **6.** 48, 1152, 24; 18; 64; 48, 24, 1152

Page 34 **1.** 15, 945, 63; 27; 35; 21, 45, 945 **2.** 8, 336, 42; 24; 14; 28, 12, 336 **3.** 72, 648, 9; 24; 27; 24, 27, 648 **4.** 28, 672, 24; 16; 42; 24, 28, 672 **5.** 48, 3456, 72; 36; 96; 32, 108, 3456 **6.** 40, 720, 18; 30; 24; 15, 48, 720

Page 35 **1.** 20, 1080, 54; 30; 36; 45, 24, 1080 **2.** 24, 576, 24; 36; 16; 96, 6, 576 **3.** 63, 2646, 42; 49; 54; 42, 63, 2646 **4.** 10, 120, 12; 20; 6; 15, 8, 120 **5.** 44, 792, 18; 99; 8; 22, 36, 792 **6.** 42, 1680, 40; 56; 30; 35, 48, 1680

Page 36 **1.** 20, 240, 12; 40; 6; 30, 8, 240 **2.** 45, 1890, 42; 54; 35; 63, 30, 1890 **3.** 40, 1440, 36; 45; 32;

20, 72, 1440 **4.** 63, 1134, 18; 63; 18; 14, 81, 1134 **5.** 48, 960, 20; 24; 40; 30, 32, 960 **6.** 48, 1680, 35; 40; 42; 56, 30, 1680

Page 37 **1.** 6, 252, 42; 14; 18; 12, 21, 252 **2.** 40, 2240, 56; 70; 32; 80, 28, 2240 **3.** 6, 270, 45; 27; 10; 15, 18, 270 **4.** 24, 1296, 54; 72; 18; 108, 12, 1296 **5.** 21, 1344, 64; 24; 56; 24, 56, 1344 **6.** 24, 2376, 99; 36; 66; 44, 54, 2376

Page 38 **1.** 40, 640, 16; 80; 8; 20, 32, 640 **2.** 6, 324, 54; 27; 12; 18, 18, 324 **3.** 63, 2205, 35; 49; 45; 35, 63, 2205 **4.** 48, 1008, 21; 18; 56; 42, 24, 1008 **5.** 15, 675, 45; 45; 15; 25, 27, 675 **6.** 56, 448, 8; 14; 32; 28, 16, 448

Page 39 **1.** 56, 336, 6; 24; 14; 16, 21, 336 **2.** 24, 1080, 45; 54; 20; 30, 36, 1080 **3.** 12, 672, 56; 16; 42; 14, 48, 672 **4.** 27, 540, 20; 15; 36; 12, 45, 540 **5.** 14, 336, 24; 12; 28; 8, 42, 336 **6.** 36, 1620, 45; 54; 30; 30, 54, 1620

Page 40 **1.** 24, 336, 14; 6; 56; 21, 16, 336 **2.** 28, 1400, 50; 35; 40; 70, 20, 1400 **3.** 16, 432, 27; 24; 18; 72, 6, 432 **4.** 56, 3024, 54; 48; 63; 72, 42, 3024 **5.** 35, 1050, 30; 35; 30; 42, 25, 1050 **6.** 27, 756, 28; 36; 21; 63, 12, 756

Page 41 **1.** $14 = 1 \times 14 = 2 \times 7$; 1,2,7,14 **2.** $27 = 1 \times 27 = 3 \times 9$; 1,3,9,27 **3.** $45 = 1 \times 45 = 3 \times 15 = 5 \times 9$; 1,3,5,9,15, 45 **4.** $24 = 1 \times 24 = 2 \times 12 = 3 \times 8 = 4 \times 6$; 1,2,3,4,6,8,12,24 **5.** $35 = 1 \times 35 = 5 \times 7$; 1,5,7,35 **6.** $22 = 1 \times 22 = 2 \times 11$; 1,2,11,22 **7.** $16 = 1 \times 16 = 2 \times 8 = 4 \times 4$; 1,2,4,8,16 **8.** $30 = 1 \times 30 = 2 \times 15 = 3 \times 10 = 5 \times 6$; 1,2,3,5,6,10,15,30 **9.** 1 **10.** 8 **11.** 2 **12.** 1 **13.** 1 **14.** 1 **15.** 9 **16.** 15

Page 42 Starting with the top row of circles, missing numbers are given row by row from left to right. **1.** 35, 840, 24; 6; 4; 28, 30, 840 **2.** 21, 378, 18; 3; 2; 63, 378 **3.** 576; 32; 3, 6, 18; 24, 24, 576 **4.** 528, 8; 4; 6, 12; 22, 24, 528 **5.** 1050, 35; 3, 7, 21; 50; 15, 1050 **6.** 48, 1152, 24; 6; 8; 48, 24, 1152

Page 43 **1.** $25 = 1 \times 25 = 5 \times 5$; 1,5,25 **2.** $46 = 1 \times 46 = 2 \times 23$; 1,2,23,46 **3.** $32 = 1 \times 32 = 2 \times 16 = 4 \times 8$; 1,2,4,8,16, 32 **4.** $36 = 1 \times 36 = 2 \times 18 = 3 \times 12 = 4 \times 9 = 6 \times 6$; 1,2,3,4,6, 9,12,18,36 **5.** $39 = 1 \times 39 = 3 \times 13$; 1,3,13,39 **6.** $62 = 1 \times 62 = 2 \times 31$; 1,2,31,62 **7.** $44 = 1 \times 44 = 2 \times 22 = 4 \times 11$; 1,2,4,11,22,44 **8.** $54 = 1 \times 54 = 2 \times 27 = 3 \times 18 = 6 \times 9$; 1,2,3,6,9,18,27,54 **9.** 4 **10.** 4 **11.** 3 **12.** 2 **13.** 2 **14.** 18 **15.** 1 **16.** 1

Page 44 Starting with the top row of circles, missing numbers are given row by row from left to right. **1.** 945, 63; 9; 5, 35; 21, 45, 945 **2.** 336; 24; 2, 14; 28, 12, 336 **3.** 72, 648, 9; 3; 3, 27; 27, 648 **4.** 72, 1512; 7; 8, 24; 27, 1512 **5.** 15, 1080, 72; 3, 27; 8; 45, 1080 **6.** 420; 2, 5, 10; 6; 42; 12, 420

Page 45 **1.** $15 = 1 \times 15 = 3 \times 5$; 1,3,5,15 **2.** $6 = 1 \times 6 = 2 \times 3$; 1,2,3,6 **3.** $12 = 1 \times 12 = 2 \times 6 = 3 \times 4$; 1,2,3,4,6,12 **4.** $40 = 1 \times 40 = 2 \times 20 = 4 \times 10 = 5 \times 8$; 1,2,4,5,8,10,20,40 **5.** $26 = 1 \times 26 = 2 \times 13$; 1,2,13,26 **6.** $13 = 1 \times 13$; 1,13 **7.** $18 = 1 \times 18 = 2 \times 9 = 3 \times 6$; 1,2,3,6,9,18 **8.** $42 = 1 \times 42 = 2 \times 21 = 3 \times 14 = 6 \times 7$; 1,2,3,6,7,14,21,42 **9.** 3 **10.** 2 **11.** 1 **12.** 5 **13.** 6 **14.** 13 **15.** 6 **16.** 2

Page 46 Starting with the top row of circles, missing numbers are given row by row from left to right. **1.** 20, 1080, 54; 5; 9, 4, 36; 24 **2.** 24; 36; 8, 2, 16; 96, 6, 576 **3.** 2646; 49; 6, 9, 54; 42, 63, 2646 **4.** 12; 3, 15; 2, 8; 20, 6, 120 **5.** 792; 11, 9, 99; 2, 8; 792 **6.** 1680; 56; 6, 5, 30; 42, 40, 1680

Page 47   **1.** 35=1×35=5×7; 1,5,7,35   **2.** 8=1×8= 2×4; 1,2,4,8   **3.** 28=1×28=2×14=4×7; 1,2,4,7,14,28   **4.** 48=1×48=2×24=3×16=4×12=6×8; 1,2,3,4,6,8, 12,16,24,48   **5.** 49=1×49=7×7; 1,7,49   **6.** 21=1×21= 3×7; 1,3,7,21   **7.** 20=1×20=2×10=4×5; 1,2,4,5,10, 20   **8.** 42=1×42=2×21=3×14=6×7; 1,2,3,6,7,14,21, 42   **9.** 7   **10.** 21   **11.** 4   **12.** 7   **13.** 4   **14.** 7   **15.** 5   **16.** 2

Page 48   Starting with the top row of circles, missing numbers are given row by row from left to right.   **1.** 20, 240, 12; 10, 40; 2, 6; 30   **2.** 1890; 9, 54; 7, 35; 63, 30, 1890   **3.** 40, 1440, 36; 9; 8; 20, 72, 1440   **4.** 18, 1134; 3, 27; 7, 42; 21, 54   **5.** 48, 3024, 63; 9; 7; 42, 72, 3024   **6.** 630; 3, 5, 15; 6, 42; 18, 630

Page 49   **1.** 23=1×23; 1, 23   **2.** 14=1×14=2×7; 1, 2,7,14   **3.** 50=1×50=2×25=5×10; 1,2,5,10,25,50   **4.** 56=1×56=2×28=4×14=7×8; 1,2,4,7,8,14,28,56   **5.** 46=1×46=2×23; 1,2,23,46   **6.** 8=1×8=2×4; 1, 4,8   **7.** 52=1×52=2×26=4×13; 1,2,4,13,26,52   **8.** 75=1×75=3×25=5×15; 1,3,5,15,25,75   **9.** 4   **10.** 25   **11.** 2   **12.** 2   **13.** 23   **14.** 2   **15.** 1   **16.** 14

Page 50   Starting with the top row of circles, missing numbers are given row by row from left to right.   **1** 576, 24; 12, 3; 16; 96, 576   **2.** 48, 3024; 54; 7, 56; 42, 72   **3.** 66, 8; 11, 44; 6; 24, 528   **4.** 1680; 8, 56; 5, 30; 42, 40, 1680   **5.** 30, 1050; 3; 5, 50; 15, 70, 1050   **6.** 40, 1440, 36; 9; 4, 32; 72, 1440

Page 51   **1.** 1   **2.** 1,2   **3.** 1,3   **4.** 1,2,4   **5.** 1,5   **6.** 1,2, 3,6   **7.** 1,7   **8.** 1,2,4,8   **9.** 1,3,9   **10.** 1,2,5,10   **11.** 1, 11   **12.** 1,2,3,4,6,12   **13.** 1,13   **14.** 1,2,7,14   **15.** 1,3,5, 15   **16.** 1,2,4,8,16   **17.** 1,17   **18.** 1,2,3,6,9,18   **19.** 1,19   **20.** 1,2,4,5,10,20   **21.** 1,3,7,21   **22.** 1,2,11, 22   **23.** 1, 23   **24.** 1,2,3,4,6,8,12,24   **25.** 1,5,25   **26.** 1,2,13,26   **27.** 1, 3,9,27   **28.** 1,2,4,7,14,28   **29.** 1,29   **30.** 1,2,3,5,6,10,15, 30   **31.** 1,31.   **32.** 1,2,4,8,16,32

Page 52   Starting with the top row of circles, missing numbers are given row by row from left to right.   **1.** 42; 7, 14; 3, 18; 12, 21, 252   **2.** 2240, 56; 10, 7; 4, 32; 80, 28   **3.** 270, 45; 3, 9, 27; 2; 15, 18   **4.** 84, 4536; 12; 63; 108, 42, 4536   **5.** 1344; 8, 24; 7, 56; 24, 56, 1344   **6.** 24, 2376, 99; 9; 6; 44, 54, 2376

Page 53   **1.** 1,3,11,33   **2.** 1,2,17,34   **3.** 1,5,7,35   **4.** 1,2,3,4,6,9,12,18,36   **5.** 1,37   **6.** 1,2,19,38   **7.** 1,3, 13,39   **8.** 1,2,4,5,8,10,20,40   **9.** 1, 41   **10.** 1,2,3,6,7, 14,21,42   **11.** 1,43   **12.** 1,2,4,11,22,44   **13.** 1,3,5,9,15, 45   **14.** 1,2,23,46   **15.** 1,47   **16.** 1,2,3,4,6,8,12,16,24 48   **17.** 1,7,49   **18.** 1,2,5,10,25,50   **19.** 1,3,17,51   **20.** 1,2,4,13,26,52   **21.** 1,53   **22.** 1,2,3,6,9,18,27,54   **23.** 1,5,11,55   **24.** 1,2,4,7,8,14,28,56   **25.** 1,3,19,57   **26.** 1,2,29,58   **27.** 1,59   **28.** 1,2,3,4,5,6,10,12,15,20,30, 60   **29.** 1,61   **30.** 1,2,31,62   **31.** 1,3,7,9,21,63   **32.** 1,2, 4,8,16,32,64

Page 54   Starting with the top row of circles, missing numbers are given row by row from left to right.   **1.** 16; 10, 8,80; 4, 8; 32, 640   **2.** 54; 9; 6, 2, 12; 18, 18, 324   **3.** 2205; 7, 7, 49; 45; 35, 63, 2205   **4.** 21, 1008, 3, 6; 7, 56; 24, 1008   **5.** 2688, 64; 7; 6, 48; 56, 48, 2688   **6.** 72; 3, 24; 9, 63; 56, 27, 1512

Page 55   **1.** 1,5,13,65   **2.** 1,2,3,6,11,22,33,66   **3.** 1, 67   **4.** 1,2,4,17,34,68   **5.** 1,3,23,69   **6.** 1,2,5,7,10,14, 35, 70   **7.** 1,71   **8.** 1,2,3,4,6,8,9,12,18,24,36,72   **9.** 1, 73   **10.** 1,2,37,74   **11.** 1,3,5,15,25,75   **12.** 1,2,4,19,38, 76   **13.** 1,7,11,77   **14.** 1,2,3,6,13,26,39,78   **15.** 1,79

16. 1,2,4,5,8,10,16,20,40,80   **17.** 1,3,9,27,81   **18.** 1,2, 41,82   **19.** 1,83   **20.** 1,2,3,4,6,7,12,14,21,28,42,84   **21.** 1,5,17,85   **22.** 1,2,43,86   **23.** 1,3,29,87   **24.** 1,2,4, 8,11,22,44,88   **25.** 1,89   **26.** 1,2,3,5,6,9,10,15,18,30, 45,90   **27.** 1,7,13,91   **28.** 1,2,4,23,46,92   **29.** 1,3,31,93   **30.** 1,2,47,94   **31.** 1,5,19,95   **32.** 1,2,3,4,6,8,12,16,24, 32,48,96

Page 56   Starting with the top row of circles, missing numbers are given row by row from left to right.   **1.** 27, 756, 28; 3, 7, 21; 4; 63   **2.** 56, 448, 8; 7, 2, 14; 4; 16   **3.** 40, 56; 10; 8, 4, 32; 28, 2240   **4.** 44, 792; 11; 4, 8; 22, 792   **5.** 1134; 3, 9, 27; 21, 42; 189   **6.** 40, 1440, 36; 9; 8; 20, 72, 1440

Page 57   **1.** 3   **2.** 8   **3.** 12   **4.** 7   **5.** 17   **6.** 15   **7.** 2   **8.** 26   **9.** 31   **10.** 1   **11.** 36   **12.** 2   **13.** 9   **14.** 9   **15.** 5   **16.** 2   **17.** 11   **18.** 14   **19.** 17   **20.** 22

Page 58   Starting with the top row of circles, missing numbers are given row by row from left to right.   **1.** 675, 45; 5, 9, 45; 5, 15; 27   **2.** 24, 2376, 99; 4; 6, 66; 44, 2376   **3.** 1296, 54; 6; 9, 2, 18; 12, 1296   **4.** 42, 1680; 7, 8, 56; 30; 48, 1680   **5.** 48, 960, 20; 4, 24; 5, 8; 30   **6.** 1080, 54; 6; 4, 36; 45, 1080

Page 59   **1.** 9   **2.** 10   **3.** 7   **4.** 32   **5.** 17   **6.** 28   **7.** 1   **8.** 29   **9.** 16   **10.** 9   **11.** 1   **12.** 29   **13.** 16   **14.** 46   **15.** 17   **16.** 18   **17.** 25   **18.** 3   **19.** 31   **20.** 1   **21.** 2, 3, 5, 7, 11, 13, 17, 19, 23, 29, 31, 37, 41, 43, 47, 53, 59, 61, 67, 71, 73, 79, 83, 89, 97

Page 60   Starting with the top row of circles, missing numbers are given row by row from left to right.   **1.** 2646, 42; 7, 7; 9; 42, 63, 2646   **2.** 45, 1890, 42; 9, 6; 5, 35; 30   **3.** 1134, 18; 7, 9, 63; 9; 14, 81   **4.** 48, 1152, 24; 3; 8; 48, 24, 1152   **5.** 336, 42; 4, 6; 2, 14; 28, 12   **6.** 1080; 6, 30; 4, 36; 45, 24, 1080

Page 61   **1.** 2×3   **2.** 2×5   **3.** 2×7   **4.** 3×5   **5.** 3×7   **6.** 5×7   **7.** 2×31   **8.** 2×23   **9.** 3×13   **10.** 5×11   **11.** 2×17   **12.** 3×19   **13.** 3×23   **14.** 5×13   **15.** 2×3× 5   **16.** 2×3×11   **17.** 2×5×7   **18.** 2×3×7   **19.** 2×3×13   **20.** 3×5×11

Page 62   **1.** 3, 2, 3 to the second power, 9   **2.** 5, 3, 5 to the third power, 125   **3.** 2, 5, 2 to the fifth power, 32   **4.** 4, 3, 4 to the third power, 64   **5.** 5, 4, 5 to the fourth power, 625   **6.** 10, 5, 10 to the fifth power, 100,000   **7.** 11, 2, 11 to the second power, 121   **8.** 21, 3, 21 to the third power, 9261

Page 63   **1.** 3×13   **2.** 5×19   **3.** 5×23   **4.** 3×41   **5.** 7×23   **6.** 2×37   **7.** 2×41   **8.** 3×53   **9.** 11×13   **10.** 2×7×11   **11.** 3×61   **12.** 3×37   **13.** 5×41   **14.** 3×71   **15.** 5×29   **16.** 5×7×11   **17.** 3×11×17   **18.** 2×7×17   **19.** 2×5×17   **20.** 3×5×23

Page 64   **1.** 4   **2.** 8   **3.** 16   **4.** 32   **5.** 64   **6.** 9   **7.** 27   **8.** 81   **9.** 16   **10.** 64   **11.** 49   **12.** 343   **13.** 64   **14.** 512   **15.** 81   **16.** 100   **17.** 1000   **18.** 10,000   **19.** 100,000   **20.** 121

Page 65   **1.** 3×17   **2.** 7×19   **3.** 3×5×11   **4.** 5×43   **5.** 3×53   **6.** 7×17   **7.** 7×23   **8.** 2×3×23   **9.** 7×31   **10.** 5×37   **11.** 7×43   **12.** 3×5×17   **13.** 2×3×43   **14.** 3×7×13   **15.** 2×5×17   **16.** 3×11×13   **17.** 2×3× 19   **18.** 2×5×23   **19.** 5×7×17   **20.** 11×41

Page 66   **1.** 8   **2.** 81   **3.** 25   **4.** 125   **5.** 36   **6.** 216   **7.** 343   **8.** 64   **9.** 81   **10.** 729   **11.** 100   **12.** 1,000   **13.** 10,000   **14.** 18   **15.** 24   **16.** 45   **17.** 175   **18.** 250   **19.** 72   **20.** 100

Page 67 **1.** 2×3×19 **2.** 5×7×13 **3.** 3×7×13 **4.** 2×5×23 **5.** 2×3×29 **6.** 2×13×19 **7.** 3×5×7×7 **8.** 2×3×23 **9.** 3×5×11 **10.** 3×7×11 **11.** 5×7×11 **12.** 2×5×3×11 **13.** 2×2×2×107 **14.** 3×13×17 **15.** 3×7×17 **16.** 2×3×11×23 **17.** 5×11×13 **18.** 5×11×19 **19.** 7×11×17 **20.** 5×7×31

Page 68 **1.** 27 **2.** 16 **3.** 125 **4.** 49 **5.** 64 **6.** 729 **7.** 10,000 **8.** 121 **9.** 1728 **10.** 169 **11.** 150 **12.** 693 **13.** 800 **14.** 4225 **15.** 1568 **16.** 756 **17.** 196 **18.** 180 **19.** 1323 **20.** 20,449

Page 69 **1.** 3×17 **2.** 5×19 **3.** 5×23 **4.** 2×5×23 **5.** 3×53 **6.** 2×37 **7.** 7×23 **8.** 2×79 **9.** 3×5×11 **10.** 3×3×17 **11.** 3×61 **12.** 2×3×43 **13.** 2×2×26 **14.** 3×71 **15.** 5×29 **16.** 3×11×13 **17.** 3×11×17 **18.** 3×7×11 **19.** 2×5×17 **20.** 3×5×23

Page 70 **1.** 8 **2.** 32 **3.** 9 **4.** 25 **5.** 125 **6.** 49 **7.** 121 **8.** 169 **9.** 289 **10.** 361 **11.** 90 **12.** 735 **13.** 120 **14.** 675 **15.** 676 **16.** 847 **17.** 425 **18.** 464 **19.** 1125 **20.** 567

Page 71 **1.** 21,280 **2.** 24,420 **3.** 30,400 **4.** 405,600 **5.** 360,270 **6.** 81 **7.** 4578 **8.** 2128 **9.** 32,500 **10.** 47,656 **11.** 18,000 **12.** 27,081

Page 72 **1.** 13,590 **2.** 37,660 **3.** 96,000 **4.** 35,538 **5.** 24,969 **6.** 328,984 **7.** 528 **8.** 1085 **9.** 1075 **10.** 899 **11.** 2268 **12.** 1054

Page 73 **1.** 25,792 **2.** 35,784 **3.** 54,459 **4.** 78,694 **5.** 184,805 **6.** 295,072 **7.** 105,075 **8.** 82,289 **9.** 147,212 **10.** 190,994 **11.** 202,300 **12.** 227,354

Page 74 **1.** 183,366 **2.** 213,460 **3.** 738,153 **4.** 330,048 **5.** 354,105 **6.** 394,212 **7.** 244,496 **8.** 202,831 **9.** 95,832 **10.** 391,773 **11.** 119,746 **12.** 530,472

Page 75 **1.** 117,691 **2.** 141,532 **3.** 140,070 **4.** 262,515 **5.** 227,916 **6.** 75,000 **7.** 403,865 **8.** 663,984 **9.** 334,971 **10.** 243,150 **11.** 317,350 **12.** 660,060

Page 76 **1.** $175 **2.** 89,760 feet **3.** 495 miles **4.** 484 words **5.** 5475 days **6.** 900 inches

Page 77 **1.** 101,360 **2.** 69,806 **3.** 113,133 **4.** 260,145 **5.** 497,426 **6.** 847,695 **7.** 378,626 **8.** 141,190 **9.** 413,748 **10.** 280,665 **11.** 603,104 **12.** 588,804

Page 78 **1.** $325 **2.** 72,600 centimeters **3.** 5000 miles **4.** 390 books **5.** 884 weeks **6.** 35,000 meters

Page 79 **1.** 66,456 **2.** 101,300 **3.** 272,362 **4.** 101,232 **5.** 260,320 **6.** 69,734 **7.** 355,756 **8.** 270,215 **9.** 348,640 **10.** 272,349 **11.** 436,278 **12.** 246,568

Page 80 **1.** $∠15 **2.** 1800 inches **3.** 624 miles **4.** 1728 pencils **5.** 4380 days **6.** 3520 decimeters

Page 81 2,3,5,7,11,13,17,19,23,29,31,37,41,43,47 **1.** 2²×7 **2.** 5×17 **3.** 2²×3×5 **4.** 2⁴×7 **5.** 5×23 **6.** 3×61 **7.** 2×5×19 **8.** 5²×7² **9.** 2³×5×7 **10.** 3×5×17 **11.** 2⁵×3 **12.** 13² **13.** 2²×67 **14.** 2³×23 **15.** 2×71 **16.** 3²×31 **17.** 2²×5×13 **18.** 2×7×13 **19.** 2³×3×19 **20.** 2²×47

Page 82 **1.** 185 **2.** 662 **3.** 829 **4.** 998 **5.** 474 **6.** 508 **7.** 2620 **8.** 5209 **9.** 7202 **10.** 6200 **11.** __ __ 58 **12.** __ 406 **13.** __ 165 **14.** 8910

Page 83 2,3,5,7,11,13,17,19,23,29,31,37,41,43,47 **1.** 2²×5 **2.** 3²×7 **3.** 2³×5 **4.** 3×5×7 **5.** 2³×13 **6.** 3²×17 **7.** 5²×7 **8.** 7×11×13 **9.** 2×3²×17 **10.** 2⁵×11 **11.** 2⁴×5 **12.** 3³×7 **13.** 2³×31 **14.** 2³×17 **15.** 2²×5×7 **16.** 3×5×17 **17.** 2×5³ **18.** 2×5×19 **19.** 2⁴×31 **20.** 11×17

Page 84 **1.** 157 **2.** 432 **3.** 575 **4.** 881 **5.** 726 **6.** 407 **7.** 1330 **8.** 4117 **9.** 3615 **10.** 4800 **11.** __ __ 47 **12.** __ __ 39 **13.** __ 246 **14.** 6112

Page 85 2,3,5,7,11,13,17,19,23,29,31,37,41,43,47 **1.** 5²×23 **2.** 3²×7 **3.** 5×7×13 **4.** 2²×3²×29 **5.** 2³×43 **6.** 3³×37 **7.** 2⁴×5×13 **8.** 7²×41 **9.** 5²×19 **10.** 2³×31 **11.** 2²×47 **12.** 3³×7 **13.** 2×5²×29 **14.** 7²×13 **15.** 2²×61 **16.** 3²×19 **17.** 2×5×37 **18.** 2²×3×17 **19.** 5²×7 **20.** 2⁴×13

Page 86 **1.** 182 **2.** 376 **3.** 403 **4.** 625 **5.** 434 **6.** 321 **7.** 1690 **8.** 4096 **9.** 6117 **10.** 2930 **11.** __ __ 75 **12.** __ __ 68 **13.** __ 102 **14.** 1043

Page 87 2,3,5,7,11,13,17,19,23,29,31,37,41,43,47 **1.** 2²×3×11 **2.** 3×5×41 **3.** 2×3×37 **4.** 2³×37 **5.** 3²×13 **6.** 2²×7×29 **7.** 2×3×47 **8.** 2×3²×23 **9.** 2×5²×17 **10.** 3²×13 **11.** 2×3×41 **12.** 3×5×29 **13.** 2³×5×11 **14.** 2⁵×3×5 **15.** 2⁷ **16.** 3²×37 **17.** 2×3×19 **18.** 2⁵×3×5 **19.** 2²×5×7 **20.** 3⁵

Page 88 **1.** 194 **2.** 297 **3.** 267 **4.** 809 **5.** 726 **6.** 603 **7.** 1730 **8.** 3612 **9.** 3092 **10.** 6110 **11.** __ __99 **12.** __ 407 **13.** 4302 **14.** 9076

Page 89 2,3,5,7,11,13,17,19,23,29,31,37,41,43,47 **1.** 2³×7 **2.** 5²×13 **3.** 2×3²×5 **4.** 2³×11 **5.** 7²×23 **6.** 3⁵ **7.** 2×3×47 **8.** 2×5³ **9.** 11²×13 **10.** 2×3²×5 **11.** 2×3×5² **12.** 7×31 **13.** 3×5×11 **14.** 2×7×13 **15.** 5×7×11 **16.** 2⁴×13 **17.** 3²×7² **18.** 2³×17 **19.** 5²×31 **20.** 2²×5×41

Page 90 **1.** 126 **2.** 432 **3.** 692 **4.** 433 **5.** 702 **6.** 821 **7.** 1880 **8.** 4117 **9.** 8220 **10.** 4041 **11.** __ __ 82 **12.** __ __47 **13.** 3014 **14.** 8011

Page 91 **1.** 342 **2.** 893 **3.** 621 **4.** 872 **5.** 6581 **6.** 6921 **7.** 39,824 **8.** 715 **9.** 293

Page 92 **1.** 52 miles per hour **2.** $1280 per month **3.** 63 cans **4.** 58 minutes **5.** $3 per person

Page 93 **1.** 964 **2.** 821 **3.** 1367 **4.** 5037 **5.** 5077 **6.** 89,352 **7.** 427 **8.** 126 **9.** 487

Page 94 **1.** 475 miles per hour **2.** $1300 per month **3.** 45 square yards **4.** 24 hours **5.** $6 per person

Page 95 **1.** 2395 **2.** 6941 **3.** 61,279 **4.** 40,792 **5.** 59,624 **6.** 40,001 **7.** 27 **8.** 33 **9.** 65

Page 96 **1.** 15 feet **2.** 500 meters per minute **3.** 61 pounds **4.** 21 miles per gallon **5.** $1232 per month

Page 97 **1.** 83,245 **2.** 80,017 **3.** 83,157 **4.** 83 **5.** 69 **6.** 73 **7.** 813 **8.** 476 **9.** 287

Page 98 **1.** 42 pounds **2.** $1157 per month **3.** 18 feet **4.** 640 acres per section **5.** 46 miles per hour

Page 99 **1.** 4312 **2.** 3218 **3.** 3821 **4.** 4692 **5.** 46 **6.** 75 **7.** 42 **8.** 651 **9.** 437

Page 100 **1.** 30 gallons **2.** 418 miles per hour **3.** 131 people per square mile **4.** $1280 per month **5.** 500 gallons per day **6.** 125 gallons per day per person